Representative Government and Environmental Management

Representative
Government

Published for
Resources for the Future, Inc.
by The Johns Hopkins University Press, Baltimore and London

and Environmental Management

by EDWIN T. HAEFELE

Resources for the Future is a nonprofit corporation for research and education in the development, conservation, and use of natural resources and the improvement of the quality of the environment. It was established in 1952 with the cooperation of the Ford Foundation. Part of the work of Resources for the Future is carried out by its resident staff; part is supported by grants to universities and other nonprofit organizations. Unless otherwise stated, interpretations and conclusions in RFF publications are those of the authors; the organization takes responsibility for the selection of significant subjects for study, the competence of the researchers, and their freedom of inquiry.

The journal articles and addresses which formed the basis of this book were originally produced as part of RFF's studies on the quality of the environment, directed by Allen V. Kneese. Edwin T. Haefele is now director of RFF's program of regional and urban studies. The figures were drawn by Frank and Clare Ford. The book was edited by Margaret Ingram.

RFF editors: Mark Reinsberg, Joan R. Tron, Ruth B. Haas, Margaret Ingram

The Johns Hopkins University Press, Baltimore, Maryland 21218
The Johns Hopkins University Press Ltd., London

Library of Congress Catalog Number 73-16106
ISBN 0-8018-1571-1

Library of Congress Cataloging in Publication data
will be found on the last printed page of this book.

Foreword

This book is one Resources for the Future is particularly proud to present. It takes us into new dimensions of problems we have long been dealing with along other lines. Mr. Haefele is a thoroughly modern eighteenth-century man in that he brings to bear new techniques of political analysis with an undisguised preference for making the government of the American Republic as contemplated by the founding fathers really work for the benefit of the people. If the era of executive ascendancy is now drawing to a close, whether because of internal structural disabilities or because of misbehavior on the part of some leading players, then this is indeed a timely book.

Edwin T. Haefele begins by reminding us that the founders of the American Republic were in full possession of the political experience of seventeenth-century England. The government framed in Independence Hall rested firmly on the great truth forged in that earlier political struggle —that representative legislatures provide the surest way to order the deep divisions among men into viable social choices. The founders built well, and left behind them the tools to keep the structure in good repair.

Through the years, he argues, we have forgotten our political roots— first in our nineteenth-century enthusiasm for popular democracy, then in our twentieth-century enchantment with business efficiency. The states, their powers crippled, no longer work as independent political units. Our political boundaries, valid a hundred years ago, make little sense today. In the cause of efficiency, more and more power has passed to executive officials and agencies. The legislatures themselves have ceased to be genuinely representative, because of confused or fragmented jurisdictions, because of the pressure of special-interest groups, and, at the national level because of the imbalance created by rigid committee control, seniority rule, and the power of senators from one-party or under-populated states. As the author observes, Madison's grand trading arena has dwindled to a number of small and petty trading guilds, and many of the real choices are being made in the bureaucrat's office.

Focusing on the question of environmental management, Haefele carefully examines how we make social choices today and how the Constitution says we should, and presents compelling arguments for bringing the two processes back into congruence. When we attempt to take such important social issues as environmental problems "out of the political arena," when we rely at every level on executive decision rather than legislative choice, he argues, we do violence to our system of government. A single-purpose executive agency is not in the right position to judge among conflicting interests of public significance; a general-purpose elected representative is. He must and should make such judgments as a part of his primary responsibility. Such irreversible, far-reaching, and interdependent decisions as those affecting common-property resources, for example, must be put back into a legislative context, where bargaining, vote trading, and accommodation across a broad range of issues can shape policies as the founders meant them to be shaped.

Haefele suggests that if we are to solve the complicated problems we face, we will have to revive, at levels appropriate to those problems, the original concept of general-purpose legislatures, where all programs competing for a share of the public purse can be weighed and judged by men responsive to a broad constituency of voters. To accomplish this we will have to give hard thought to some thorny questions: redrawing political boundaries to capture the issues; equity versus efficiency; the present separation of substantive legislation from appropriation, and of both of them from taxation; reckoning public costs in private calculations; controlling certain excesses of interest groups; and perhaps more profound changes in the formal elements of the political system. Haefele proposes a return to constitutional government as the best way for dealing with these issues rather than continuing to rely so heavily on executive and bureaucratic means.

The papers in this volume have been brought together from several

earlier ones published in professional journals, somewhat rewritten, and
provided with connecting tissue. Because the issues addressed are timely
and the ideas Haefele works with are seminal, it seemed desirable to make
the collection available at this time. Resources for the Future hopes to
continue probing these matters in an effort to throw more light on ways
to improve the functioning and the responsiveness of representative
government in this country.

JOSEPH L. FISHER, *President*

Resources for the Future, Inc.

August 1973

Contents

Tables

Appendix Tables

Figures

Appendix Figures

Representative Government and
Environmental Management

1

1976 and the Machinery of American Government

There will be many commemorations of the 200th anniversary of American independence and it is fitting that there should be. The Republic is now, as it long has been, "the best hope of mankind" and no cavil about our present problems diminishes that hope one whit. Indeed, our problems persist because we persist in trying to achieve the goals set down 200 years ago by pragmatic men who were confident enough to strive for excellence.

It would be, I believe, a disappointment to those men to know that we have left out one element in our plans of commemoration, self-praise, and celebration. As men consciously constructing the machinery of self-government, they were aware they built imperfectly and they depended upon later generations to change and perfect as experience demanded. George Washington wrote to his nephew Bushrod Washington on November 10, 1787, as follows:

> The warmest friends and the best supporters the Constitution has do not contend that it is free from imperfections; but they found them unavoidable and are sensible, if evil is likely to arise therefrom, the remedy must come hereafter; for in the present moment it is not to be obtained; and, as there is a Constitutional door open for it, I think the people (*for it is with them to judge*) can, as they will have the advantage of experience on their side, decide with as much propriety on the alterations and amendments which are necessary, as ourselves. I do not think we are inspired, have more wisdom or possess more virtue than those who will come after us.

It is worth noting that eleven of the twenty-four amendments to the Constitution were passed before 1800. As we approach our 200th anniversary, the words of Washington have a hollow ring, because we regularly teach our children about the inspiration and wisdom of the founding fathers, and our high school civics books (the only exposure most people get to the theory of American government) espouse a faith in our institutions rather than teach an understanding of the principles behind them. It is no wonder that, for the most part, Americans are afraid to consider changes in basic governmental form. What we do not understand we are afraid to touch. What we have been taught to revere we are not likely to want to change.

Some in the generation of the sixties took the opposite path. Rejecting the teachings of faith because they did not square with the facts of undeclared war, poverty, racial misery, and environmental degradation, these people also in large part rejected American governmental institutions *in toto*.

Our 200th anniversary gives us the opportunity to correct both the attitude of unquestioning faith in our institutions, which would strike the founders of the Republic as unjustified, and the attitude of unquestioning rejection of our institutions, which would strike the founders as irrational. We could, as a part of our celebrations, make a conscious and dedicated attempt to relearn the lessons of our constitution, most of which are in seventeenth-century English history, a history that both Jefferson and Adams constantly recalled to their countrymen. What Americans today, other than historians, know the first thing about those constitutional struggles in England out of which this Whig Republic was created? Who could now tell you why the Republic is properly called a "Whig" republic? Lest this be considered a pedantic trifle, compare the speeches in Commons of Sir John Eliot in the 1620s with those of Senator J. W. Fulbright on the subject of foreign wars fought without legislative support, and wonder how we could have forgotten so easily that executive power must be constantly checked if it is not to dominate all options and future choices.

In the course of relearning why it was we constructed the kind of government we did, we have one other obligation most appropriately to be performed as a part of our 200th anniversary; that is, to examine what changes our government needs in order for it to function as the founders intended it should. Two things are implied by the last sentence. First, that we do the most honor to the memory of the founders of this nation by trying to improve on their handiwork instead of worshiping it and, second, that it is possible to examine the fundamental mechanics of government (as distinct from the programs and administrative boxes of government) in light of its ability to perform its given tasks.

The latter concept is perhaps better illustrated than described. Take the case of legislative representation. The original purpose of having a representative assembly at state, local, and national levels was to allow citizens to control their government—to set the agenda for deliberations, to propose and pass or reject laws, and to oversee the administration of laws previously passed. It is a commonplace observation that nowadays many innovations in legislatures—rigid committee control, the seniority system for committee assignments, and the like—give some legislators much more power than others. These developments were not anticipated in the eighteenth century, because they so clearly violate the purpose of a legislative assembly. Yet, ironically, we resist changing these features of present legislatures because of "tradition" and thus venerate something our forefathers would have abhorred.

Another example may be found in the question of boundaries for governments of general jurisdiction. In the eighteenth century it was reasonably clear that state boundaries and subunits of states followed viable political communities. Nowadays it is unusual for a state, county, or city boundary to possess the slightest amount of political sense. Yet again we venerate the existing situation, unaware that the eighteenth-century mind would not have tolerated for a moment the political fragmentation that present boundaries force upon us.

One of the hardest and most untouched problems of basic structure we have is how to devise a method of dealing with appropriate boundaries. The day has long gone when we could make sense out of George Mason's criterion for suffrage: ". . . every man having sufficient evidence of attachment to and permanent common interest with the society."[1] Yet we keep a voting registration pattern that pretends a man's interest lies primarily where his house is. In any metropolitan area, with its fragmented jurisdictions, such a definition of citizenship tempts some to neglect their responsibilities and frustrates others who would exercise theirs. We ignore such breakdowns in the possibility of self-government to our ultimate peril as

[1] Debates in the Federal Convention of 1787 as reported by James Madison, August 7, 1787.

free men, for our country has always been abundant in men who would govern us if we do not govern ourselves.

A Brief Historical Sketch

On August 12, 1810, Thomas Jefferson wrote from Monticello to William Duane, editor of the political newspaper, *Aurora*: "Our law, language, religion, politics and manners are so deeply laid in English foundations that we shall never cease to consider their history as a part of ours and to study ours in [light of] its origins." Within a generation of the time he wrote, America had so turned westward that we lost all popular knowledge of the English foundations of our law and politics and began to change and extend our governmental structure in ways that did not always square with those foundations. It may be useful to look again at the foundations in order to see where and when we lost them and hence lost their supporting strength for the government we constructed.

In modern times, the story begins in Stuart England with the conflict between the executive and the legislature, the Crown and Parliament. Sir John Eliot, who was to die in the Tower, speaks in Commons during the second parliament of the reign of Charles I: "Our honour is ruined, our ships are sunk, our men perished; not by the sword, not by the enemy, not by chance, but . . . by those we trust."[2] The immediate focus of Eliot's attack was the incompetence of the king's favorite, Buckingham, but the citadel being stormed was royal prerogative. The history of England in the seventeenth century is the tale of free men taking the measure of the royal prerogative and wresting it by force of arms from two Stuart kings. When the crown was given, by Parliament, to William of Orange in 1689, there was a clear notion, on both sides, that ultimate sovereignty remained with Parliament.

Within a generation of the Glorious Revolution in England, the legislatures in most American colonies had successfully resisted the royal prerogative of the crown-appointed governors; by 1740 the colonial legislatures, like their English counterpart, were dominant in fact if not yet in law. Legislative supremacy was a fact taken for granted by the Whig generation in America. Madison observes almost parenthetically in *The Federalist*,

[2] S. R. Gardiner, *History of England* (London: Longmans, Green and Co., 1884), VI, 62.

Paper no. 51, that "in republican government, the legislative authority necessarily predominates." Even Hamilton, who was certainly no Whig, says in Paper no. 73 that "the superior weight and influence of the legislative body *in a free government* and the hazard to the Executive in a trial of strength with that body, affords a satisfactory security" (against executive dominance), with the assurance that his audience would agree.

The whole sense of the system of government that was *re*-formed in the states and in the federal union in the 1780s was rooted in the belief that representative legislatures provided the surest mechanism for ordering the deep divisions to which mankind is prone into viable social choices. The belief was not a faith in an untried theory, but a judgment based on nearly 200 years of English and American experience with bloody conflict over religion, control of foreign policy, and civil liberty. No cavil about the limited extent of the franchise should obscure the point. The judgment was based on political *history* rather than political *philosophy*, and the writings of James Harrington, Algernon Sydney, and John Locke played a less direct role than is often ascribed to them. The spirit of the Philadelphia Convention is well summarized by John Dickinson's words, "Experience must be our only guide, for reason may mislead us."[3] The American Constitution was a codification of experience, mostly English experience gained at considerable cost. It has long been recognized by legal scholars that the basic forms of American government reflect England at the close of the seventeenth century rather than the England of 1776. The distinction is critical, because, while we adopted the Whig principle of legislative supremacy for policy determination, we rejected the merging of legislative, judicial, and executive power that Walpole accomplished in eighteenth-century England and that stifles the Mother of Parliaments to this day. Instead, we kept to the older Whig principle of the independence of the common law and an independent judiciary as the interpreters of constitutional law. We also constructed two institutional innovations, an independent but accountable executive (at the federal level only; none of the state governors were allowed much independence from the legislature) and a federal senate that could fill the kind of councillor role played by the House of Lords in England. It is important to remember that neither senators nor presidents were to be popularly elected. The founders were par-

[3] Debates in the Federal Convention of 1787 as reported by James Madison, August 13, 1787.

ticularly worried about the potential for mischief inherent in the popular election of the president and had no doubt about the demagoguery such an election procedure would bring out. They also worried about the balance they had created between the president's accountability to Congress and his independence from Congress. As a safeguard, they spelled out some things the executive could not do (declare war, for example), but there was no general ease of mind about the outcome. Similarly, the system for electing the president was a complicated process, the strengths and weaknesses of which were not totally apparent to the founders.

At the urging of many, including George Mason, who would not approve the Constitution as written, a bill of individual rights was made a part of the Constitution in order to ensure that the "rights of freeborn Englishmen" would explicitly be constitutionally guaranteed in America. Thus a government was put in place.

As the nineteenth century got underway in America, both John Adams and Jefferson were concerned about legislative tyranny. Jefferson commented that "173 despots would surely be as oppressive as one."[4] The rest of the country did not discover this possibility until later in the century, when it became apparent to all that unless legislatures in the states were restricted they would steal everything loose. Since most states started with sovereign powers unlimited except by power delegated to the federal government, the power to pillage was immense. Land grabs, special-purpose legislation, railroad investment schemes, canal bonds, and other inventions for private profit at public expense abounded. In desperation, the many state constitutional conventions of the nineteenth century adopted new articles that severely limited the powers of the state legislatures to act and even more severely limited the power of the people to make further constitutional changes. It was an understandable but regrettable overreaction. Moreover, reflecting the wave of egalitarianism in continental Europe, the conventions enacted laws and constitutional requirements for referenda and other direct democracy devices and often made them mandatory when state and local taxation and bond issuances were involved. (One can leap ahead a moment to understand why the federal government, which had no such restrictions, would become the focus for new programs as they became needed. The founders had no illusions about the possibility of direct democracy.)

[4] Thomas Jefferson, *Notes on Virginia* (Boston: Lilly and Wait, 1832), p. 123.

As a part of the movement for more popular control, senators came to be elected rather than appointed in most of the states. As a result of that change, state legislatures suffered a further loss of power, and the last formal tie between the state, as a political jurisdiction, and the Congress was lost. The loss was not calculated; it was thoughtless.

America entered the twentieth century, then, with its executive governments at all levels restricted by original design, its state legislatures (the only ones with plenary powers) severely weakened by constitutional amendments, its understanding of its own system of government confused by the new enthusiasm for popular democracy, and its own history of rejection of popular democracy—in the seventeenth-century Commonwealth period in England and the eighteenth-century Revolution in America—ignored or forgotten.

In the twentieth century we began to use our business sense on the business of government. Since it has always been true that governments perform services, spend money, buy equipment, and do other things not unlike the activities of businesses, it is but a short step, in some minds at least, to judging government on business terms. Hence, at the beginning of the twentieth century, it was clear to many that government was not efficient. This view arose at about the same time (prior to and just after World War I) that we became firmly middle-class in our values and goals. "The business of America is business" may be regarded as typical of the complacency about goals and the focus on means. The rise of the city manager profession, the professionalism of the civil service, the strong mayor plans, the nonpartisan and at-large elections, and the rise of the fallacy that there is no Republican way to collect the garbage are all reflections of the tendency to view government through the eyes of the public administrator and the business executive. At the heart of all these efforts was the driving American ambition to make the system work, and a profound and growing ignorance of what the system had been designed to do.

The depression of the early thirties is a convenient bench mark for measuring the degree to which executive government (that part of government understandable on business terms) had again seized the initiative for public policy. The country turned to the presidency for leadership, as in a war. Presidential leadership was forthcoming, and the former drive for efficiency was reinforced by the need for new social programs. Federal grant-in-aid programs, which effectively put social choices at the federal level and used the states as administrative channels, became the chief device by which the federal executive initiative was maintained. Using the state and local governments as administrative channels, of course, reinforced the need for efficiency in their executive branches, and making social choices at the federal level further reduced the importance of state and local legislative bodies. World War II simply raised all of the existing tendencies and trends by an order of magnitude.

After World War II we saw a phenomenal growth in our powers to solve large-scale technical problems, and hence a further impetus to the dominance of executive governments. Executive governments have large-scale technical problems; and engineering, economics, and systems analyses disciplines developed a number of sophisticated tools—cost-benefit, cost effectiveness, operations research, planning, programming, and budgeting —to cope with the problems of executive government.

The tools and their users prospered until the middle 1960s or thereabouts, when resistance began to appear. In the highway area people began to chain themselves to bulldozers and trees. In the water resources area people being flooded out for the benefit of downstream cities began to fight in the courts. In the central city people began to resist demolition for the benefit of suburbia. The young began to resist fighting a war that benefited no one. Legislators, Senator Fulbright, for example, began to speak words that echo those of Sir John Eliot. The legitimacy of executive government, not its efficiency, was again at issue. Executive power grew, not by usurpation by any president, governor, or mayor, but from circumstances confounded by ignorance of the consequences of such power. The scholars who, during the 1930s, drew on wartime history and a one-eyed reading of legal precedents to provide the basis for the implied powers of the executive, did so as patriots, not as aspiring despots. Similarly, the constitutional amendment allowing a federal tax on personal incomes was not proposed or put into place by men who wished to destroy the states' ability to be independent fiscal units, but it has had that effect. The power of a Canadian province, which still has a large measure of fiscal independence, is striking compared to that of an American state in 1973.

These changes, in company with many other developments not recorded here, have meant that we have drifted further and further toward being a unitary state in fact while still a federal state in law. Other proposals, such as the direct election of the president, would make us even less a federal state. The implications of such changes—similar in kind to the problems of a Congress dominated by committees and by senators from one-party states and the confusion of our internal boundaries—portend a dismantling of the underlying political structure of the nation.

Yet all the changes occur without any conscious examination by Americans, who seem increasingly split into two camps—one that wants to repeat the pledge of allegiance to the flag and pretend nothing is wrong, and the other that wants to pretend that nothing is right. Neither camp could maintain an argument with even a moderately informed citizen of eighteenth-century America on the subject of the principles of our government.

A MODEST PROPOSAL FOR STUDY

The occasion of the 200th anniversary offers us an opportunity to re-dedicate ourselves to the principles of our government. No better occasion could be devised for examining the present state of governance in light of those principles. The following suggestions for study are put forth in the spirit of such rededication and examination.

1. *Redrawing the Sixteenth Amendment regarding a federal income tax.* This focuses attention on the constitutional question involved in revenue-sharing. Chairman Wilbur Mills has made a valid point in saying that present plans for revenue-sharing violate a basic principle of our government, i.e., separation of responsibility of taxation from that of spending. Yet some fundamental resolution of the fiscal imbalance between state and federal government must be accomplished if our states are to become viable again.

2. *Adding to our Bill of Rights some formulation of a citizen's right to a reasonably healthy physical environment.* This would explore the possibility of utilizing the legal system in a more fundamental way for environmental protection. Just as we do not leave freedom of speech to either legislative or executive determination, but protect it from both as a basic human right, it may be that there are elements of our environment that deserve constitutional protection.

3. *The repeal of clauses in state constitutions that restrict legislative choices.* The general criticism that state government is helpless is true, but

it is true in part because of constitutional shackles that make it impossible for legislatures to make decisions and spend money. Their removal, coupled with some solution to the federal-state problem regarding the personal income tax, could restore the states as lively and creative instruments for the making of social choices and for the management of the public business.

4. *The redrawing of state boundaries.* Hard upon the infusion of legal and fiscal strength at the state level is the necessity to rethink the state boundary question. Utilizing the two principles of Madison concerning boundaries—that is, that they encompass a heterogeneous society that has common problems—we can see that many present states do not qualify on one or both counts. Some states might well be split into two—California, for instance. Others may need to be combined or reformed into a new state. While the mere proposal to consider such changes inevitably brings out everyone from both fringes of the lunatic, it is resolutely put forth as deserving the most serious attention. It is one fundamental change that requires no constitutional amendment.

5. *Rethinking representation in the Senate.* Either as a part of redrawing state boundaries or separately, Senate representation should be used to redress the balance of interests in the nation. Questions need to be raised about urban representation, regional representation, and even national representation in the Senate. If the state pattern is drastically overhauled, then little may need to be done. If it is not, then much will have to be done. The simple fact is that the Senate cannot be kept forever in bondage by minimal population states. This was not the intention of the founders, and no rational purpose is served thereby.

6. *The present parties' hold on elective offices.* While it is impossible to force a people to govern themselves if they do not wish to, it may be that the American apathy toward its indispensable party system could be corrected if the present two parties did not have such a formal, legal lock on elective offices. Legal requirements to get on the ballot, to get into primaries, and the expense of running for office may allow an old oligarchy to hold power far beyond its popular support. By breaking these locks we might be able to resuscitate a viable two-party system.

7. *House and Senate rules.* Perhaps no tradition is more justifiably guarded in a free society than that which says that legislative bodies shall

be the arbiters of their own rules and membership. To violate that tradition goes against not just seventeenth-century history but most history of self-government in the Anglo-Saxon world. Nonetheless, if the Congress cannot, or will not, break out of the rigidity of committee rule, and if the parties will not break out of seniority as the basis of assignment to those committees, then the older first principle of self-government must break that tradition and impose rules to enable the individual legislator to function on equal terms with every other legislator.

8. *Redressing the balance between the executive and the legislative at all levels of government.* Two classes of problems plague governments at present. One class comes about because no one has authority to act. No one can manage a river basin, build low-cost housing, or reorganize executive agencies to meet new tasks. The frustrations of the powerless executive are many, varied, and dangerous. The other class of problems comes about because the executive *has* acted. Port authorities continue to make decisions adversely affecting people's lives. Highways do get built through urban communities that do not want them. We are in Southeast Asia. Actions taken by executives without legislative resolution of value conflicts are many, varied, and deadly to representative government.

Executive frustration may well be rooted in the boundary problems, both territorial and jurisdictional. Legislative impotence stems, in part, from the fact that few analytical resources are available to legislators for tracing out the distributional consequences of alternative courses of action.

9. *Decision rules in legislative bodies.* Both simple majority and two-thirds majority are used as decision rules in our legislatures, depending on the gravity of the issue at hand. We are now aware of a number of subject areas, among them land-use controls and environmental protection, where the asymmetry of the decision is pronounced—that is, where what is done cannot be undone. A wilderness cannot be recreated, nor can a coastal bay area be depopulated. Extra caution is required in taking irreversible decisions. Special majorities, passage by two different legislative sessions, and concurrence by the areas affected are all ways whereby the body politic could protect itself against hasty decisions on irreversible matters.

10. *Welfare and individual vesting.* Now that welfare programs seem on their way to becoming a federal responsibility, a major governmental prob-

lem may be resolved. Nonetheless, a number of social programs, including welfare and education, can profitably be examined through the analogy of the G.I. Bill, where the benefit vests directly to the individual, who can redeem his benefit through any public or private agency. The traditional conservative and the new left radical take almost identical stances on this issue. The present approach, maintaining large bureaucracies to "manage" social welfare programs, has few champions.

11. *Election of the president.* Although proposals for direct presidential election are misguided, some changes are needed. Four procedures make up our present electoral system: majority of electoral vote as the decision rule, award of state electoral vote on a winner-take-all basis, the use of electors to cast the state's vote, and congressional resolution of indeterminate elections. There is widespread confusion in the public mind *and in the mind of Congress* about the function of each of these procedures. For example, the national two-party system utterly depends on using a majority of the electoral vote taken state by state; yet present congressional proposals focus primarily on abolishing the electoral vote. (The electors should, of course, be abolished.) No real attention has been given to devising a congressional resolution of indeterminate elections that does not replicate the majority that caused the indeterminacy in the first place. (See appendix A.)

12. *Regulating the use of common-property resources.* It may be that present constitutional provisions can be used to levy charges on individuals and firms who use the air and water for residuals disposal. Even so, a great improvement in our ability to take this necessary action would be made by an explicit constitutional provision to this effect. Framing such a provision would provide the clearest way of examining the issues surrounding the idea of forcing private decision makers to take account of the costs they impose on other people.

THE FOCUS ON ENVIRONMENTAL QUALITY

A number of the proposals listed above are treated in some detail in later chapters. In particular, the ramifications of environmental quality in relation to representative government are explored. As it happens, these explorations lead me, inevitably, to examine the basic structure of government and its decision-making process. The question of the appropriate boundaries for governments comes up again and again. The necessity for legislative resolution of value conflict appears many times. The appropriate separation between policy and execution of policy is a recurring theme—not, it should be noted, in the old public administration sense, but in the constitutional sense.

Finally, because most of the analyses rest on a particular theory of social choice (which is reprinted in the appendices), there is a certain amount of overlap in the essays that follow. Some of it exists because each essay was originally written as a separate piece; some is simply the result of the close connection that environmental choices have to social choice. Wherever one starts, the path has a way of leading to an examination of how we make choices and how our Constitution says we should make choices. It would be reassuring if, by 1976, the two procedures should bear a closer resemblance to one another.

2
Environmental Quality as a Problem of Social Choice

INTRODUCTION

The problems posed by having to make explicit choices about the environment, something that governments now have to do, have pointed up the weakness of modern government. So long as the problems faced by governments could be defined as technical problems, then they could be passed along to experts and the solutions arrived at by a nonpolitical or covertly political process. But environmental problems quickly became value judgment problems. Moreover, they obviously require governmental action.

It was once generally true that environmental quality could be purchased in the private market. As my income rose, I could confidently look forward to enjoying cleaner air, a quieter neighborhood, and most other elements of what might have been (and was) called gracious living. Now, even though my income rises, I find the private market for environmental quality closed to all but multimillionaires, and even they are worried.

Were environmental quality still to be bought through individual transaction (like buying Cadillacs), we could ignore the social issue posed by bad environments by treating it (like ten-year-old Chevys) as an income distri-

Reprinted with minor revision from Allen V. Kneese and Blair T. Bower, eds., *Environmental Quality Analysis: Theory and Method in the Social Sciences* (Baltimore: The Johns Hopkins University Press for Resources for the Future, Inc., 1972). Copyright © 1972 by The Johns Hopkins University Press.

bution problem. When, however, rich and poor alike suffer from an environmental quality problem, we know that the invisible hand has deserted us. The effect of that desertion has turned some economists into political philosophers,[1] some biologists into polemicists,[2] and may well drive historians to despair.

Aristotle said it first: "For that which is common to the greatest number has the least care bestowed upon it."[3] Since we have not been able, in general, to assign private property rights to all the air and water, we have owned them in common and cared for them least. Now that we have overused them individually, we face the task of bringing these common-property resources into some system of governance. We shall in the future have to make collective—that is, social—choices rather than individual choices about their uses.

Problems in Making Social Choices

Making social choices brings some special problems. It is often a case of "either-or" choices, of this bundle of goods and services versus that bundle. Since public monies, like private monies, are not unlimited, spending more for A means spending less for B or C. When either-or choices are faced by individuals, each man can choose. When these choices are faced by a number of people, some of them are not going to get what they would choose. That fact has been one of the main reasons for governments throughout human history. Dictatorial and administered states view this problem differently from states that try to determine social choices from individual preferences.

Other attributes of social choices are also bothersome. Besides being either-or, they are apt to involve conflicting or multiple objectives, with no generally accepted criterion for sorting them out. When an aggregate economic efficiency measure is either implicitly or explicitly accepted as the criterion for public investment, the indeterminacy is masked. As analysts begin to probe into the incidence of costs and benefits, including externalities, the indeterminacy is once again revealed.

[1] Kenneth E. Boulding, "The Network of Interdependence" (Paper presented at the Public Choice Society, Chicago, 1970).

[2] Garrett Hardin, "The Tragedy of the Commons," *Science* 162 (December 13, 1968): 1243–48.

[3] Aristotle, *Politics*, book II, chap. 3.

Still other problems in social or collective choices relate to revealed preferences and the free-rider problem in pure public goods. If a public good is defined as a good or service, access to which cannot be denied and enjoyment of which does not diminish the quantity available to others, then it is easy to imagine situations in which it is to my profit not to state my preferences lest I be burdened with more of the cost.[4]

An equally troublesome problem closely allied to the free-rider problem is the problem of joint supply. Some common-property resources—a river system, for example—serve multiple purposes. Although it is possible to charge users of the water, be they industries, municipalities, fishermen, or boaters, it is not altogether clear what each class of users should be charged. The problem is particularly vexing if public investments are made—dams, for example, that have multiple purposes. Any true joint cost is just that, and no allocation on cost principles is possible. Lacking a cost allocation formula, authorities may price on a willingness to pay (trying for a price discrimination regime that can bring financial self-sufficiency). These schemes often run afoul of political realities—the users can influence the decision. In social choices, everyone can claim to be entitled to equity, and some to a little more equity than others.

All of the problems associated with social or collective choices were known long before economists began to be interested in public finance, welfare economics, and public goods. It has been the economist, however, who has focused the issues in modern times and who has addressed the problem with most rigor, even if often with a total lack of historical insight.

Properties of a Social Choice Mechanism

The most famous modern statement describing the formal properties of a social choice mechanism is Kenneth Arrow's.[5] Arrow laid down five (reformulated then into four) seemingly reasonable conditions that a social choice mechanism should have and found that, in general, no mechanism could be devised that met these conditions. These conditions, taken from a recent restatement by Arrow,[6] are:

[4] See Mancur Olson, *The Logic of Collective Action: Public Goods and the Theory of Groups* (Cambridge: Harvard University Press, 1965).

[5] Kenneth Arrow, *Social Choice and Individual Values*, 2nd ed. (New York: John Wiley, Inc., 1963).

[6] Kenneth Arrow, "Public and Private Values," in Sidney Hook, ed., *Human Values and Economic Policy* (New York: New York University Press, 1967), pp. 3–21.

1. Collective rationality: In any given set of individual preferences the social preferences are derivable from the individual preferences.
2. Pareto principle: If alternative *A* is preferred to alternative *B* by every single individual, then the social ordering ranks *A* above *B*.
3. Independence of irrelevant alternatives: The social choices made from any environment depend only on the preferences of individuals with respect to the alternatives in that environment.
4. Nondictatorship: There is no individual whose preferences are automatically society's preferences, independent of the preferences of other individuals.

Arrow, and others,[7] have devised restrictions on individual preferences that are sufficient to allow a social choice consistent with these conditions. The restrictions have been, however, so severe as to leave Arrow's essential theorem intact. Some have maintained that the conditions must be rejected as irrelevant to an understanding of how collective choices are actually made in a committee situation.[8]

The conditions remain, however, the clearest statement of what we might want in aggregating individual preferences into social choices. Moreover, they can be shown to be relevant when preference orderings are combined with voting stances. In another paper (see appendix C), I have shown that representative government, with a two-party system, can provide a means of going from individual choices to social choices in a way that meets all Arrow's conditions. The essence of the case is that the two-party system can function in a way that will bring out the two positions that, when voted upon, produce the same decision as is produced if the voters could have indulged in vote trading on the issues involved. I emphasize that I say representative government *could* operate as an ideal social choice mechanism, not that it does at present. This is analogous to saying that a competitive

[7] See Duncan Black, *The Theory of Committees and Elections* (Cambridge: The University Press, 1963) for the general case. See also C. R. Plott, "A Notion of Equilibrium and Its Possibility under Majority Rule," *American Economic Review* 57 (1967): 787–806; Gordon Tullock, "The General Irrelevance of the General Impossibility Theorem," *Quarterly Journal of Economics* 81 (1967): 256–270; and Gerald Kramer, "On a Class of Equilibrium Conditions for Majority Rule," mimeographed (New Haven: Cowles Foundation Paper no. 284, 1969).

[8] For example, Duncan Black, "On Arrow's Impossibility Theorem," *Journal of Law and Economics* 12 (1969): 227–248.

price structure is ideal, not to saying that current market prices are at that point.

The Present Environmental Quality Picture

The present governmental structure, which relates to environmental quality, bears little relationship to representative government, to the party system, or to social choices. Most of the choices made consciously by governments are made either by technicians who try to "balance" the interests of the affected parties or by a small group of politicians who hide their choices behind a "technical" but inadequate benefit-cost analysis. We thus have the worst of both worlds—technical analysis debased by political judgments, and political deals in which only a small number and perhaps the wrong people may play.

The criticism needs to be made explicit. I am saying two things about the technical analysis and two things about the political process. First, the technical analysis does not cover the full range of technical possibilities and, second, it is tempered by what the technician judges to be political reality. Both are grievous faults. The political process is faulty because, first, it is conducted by the wrong people (say, for example, the Public Works Committee) and, second, it is hidden behind technical surveys that purport to be objective.

It may well be that the technical and political processes now in operation served adequately in the past, perhaps in the not very distant past. Given a rather broad consensus on objectives, the focus of a problem is the determinate issue of how best to reach the objective, so long as side effects are not perceived to be important. Certainly the early success of the Port of New York Authority (PNYA) is a case in point. Only recently, when it became clear that important social choices were being made (and others foreclosed) by PNYA, has the Authority ceased to enjoy its former approval. Only when the management of our water resources involves multiple-use problems do the work of the Corps of Engineers and the Bureau of Reclamation and the deliberations of water authorities lose their purely technical character in the eyes of the public. Only recently has the origin-destination survey ceased to be accepted as the sole criterion for determining highway investment need.

In the past, technical, administrative, and executive agencies, using the device of the public hearing and citizen or special-interest advisory committees, may have been sufficient to solve the problems. We are moving now to a time when legislative government, using technical and administrative advisory committees, is needed. For when true social choices are at stake, nothing less than legislatures making these choices will suffice in our system of representative government. The spectacle of executive personnel attempting to assess the public interest through public hearings or to divine appropriate actions through committees "representing" all interests from housewives to steel mills is an outrage in the pure sense—it does violence to our system of government. In a technical sense it does not aggregate individual preferences correctly into social choices.

Paradoxically, an outrage of equal proportions is committed by the present workings of the legislative process in relation to water. With public investment in water management now largely the province of the federal government, and within Congress the province of the public works committees, something very different from a proper functioning of the legislative process occurs. The twentieth-century dominance of the seniority habit and of committee rule gives inordinate power to members from one-party states and breaks down the large-scale trading system that Madison envisioned in *The Federalist*, Paper no. 10, into a number of small trading guilds. Even more specifically unfortunate for rational water policy is the unhappy fact that water projects are a major public currency in the Congress. As one of the most visible, tangible evidences of bringing money to a state or district, water projects have been in great demand by aspiring members of Congress. Again, it may well be true that in the past water was an appropriate specie, but with multiple and conflicting water uses it becomes a far less appropriate medium of exchange. Moreover, since the water is a local or regional resource, the national Congress becomes less and less the appropriate legislative body to be making the choices.

Some Tools of Analysis

Americans were extraordinarily skilled at constructing social choice mechanisms in the seventeenth and eighteenth centuries, when the colonists used representative legislative bodies to wrest policy control from crown-appointed governors and executive councils. Later, their skill in this area seemed almost to die out. Instead, they perfected managerial and executive skills in both business and government. Faced now with value conflicts that call for social choices, we grope to relearn some of the more ancient arts. In doing so, we are helped by a number of tools developed by men in economics and related fields as they turned their attention to problems of social choice. While the present discussion is not exhaustive, it does try to identify some of the important tools, show how they relate to each other, and demonstrate that we have progressed further than is generally recognized toward having a positivistic as well as normative basis for constructing new mechanisms for making social choices.

Nominal Voting Power—The Shapley Value

In any legislature, commission, or committee composed of a number of independent members, some coalitions or blocs have always been known to be more powerful than their numbers would indicate. Different subsets of the members could be designated as large city, small city, suburban, or rural, or as industry, labor, or farm. More or less permanent protocoalitions often form around commonality of interest connoted by these designations.

Similarly, if in a two-house legislature one house has fewer members than the other, a member of the smaller house is more powerful than a member of the larger. Rousseau could thus say, with confidence, "that when the functions of government are shared by several tribunals, the less numerous [members] sooner or later acquire the greatest authority. . . ."[9]

Moreover, if weights are assigned to votes, and different entities have therefore more votes than others, the power of the different entities is not often proportional to the difference in votes. Thus, while small states have more power in the electoral college than their relative population (due to the constitutional formula of assigning electoral votes), their power is not as great as the numbers would indicate.

Sometimes the variation between the relative numbers and the real voting power is striking. Only intuitively felt in the past, the variation can now be calculated precisely, using Shapley values.[10] The Shapley value calculates

[9] Jean Jacques Rousseau, *The Social Contract*, book III, chap. 4.

[10] The formal mathematics are in L. S. Shapley, "A Value for *N*-Person Games," *Annals of Mathematics Study* 28 (1953). Application to political bodies was first made in L. S. Shapley and Martin Shubik, "A Method for Evaluating the Distribution of Power in a Committee System," *American Political Science Review* 48 (1954): 787–792. A method for exact calculation, by computer, for large bodies is in L. S. Shapley, "Values of Large Games VI: Evaluating the Electoral College Exactly," RAND–RM–3158 (Santa Monica: RAND Corporation, 1962). William Riker and L. S. Shapley explore the consequences of weighted voting in "Weighted Voting: A Mathematical Analysis for Instrumental Judgments," *Nomos* (1966).

how many times each (voter, bloc, unit, etc.) can appear in the "pivot" position by providing the winning margin for one side, assuming all possible permutations are equally likely. To give an example, assume an eleven-member group in which there are two voting blocs of three members, one two-vote bloc, and three voters who belong to no bloc. Assume also that a simple majority of six votes is needed to win on any issue. By proportion, the relative power of a three-vote bloc (3/11) would be about 27 percent, and that of a two-vote bloc and a single vote, 18 percent and 9 percent respectively. Instead, the relative strengths are more nearly 30, 20, and 7 percent, as shown in the following calculation:

A. Total arrangements possible of one two-vote bloc, two three-vote blocs, and three single votes

$$= c(6, 1) \times c(5, 3) \times c(2, 2)$$

$$= \frac{6!}{1!5!} \times \frac{5!}{2!3!} \times \frac{2!}{0!2!}$$

$$= 60 \text{ arrangements.}$$

B. 1. Number of times a single vote is a pivot (* designates pivot):

(a) 3 2 1 3 1 1
$\qquad\qquad$*

(b) 3 1 1 1 3 2
$\qquad\qquad$*

(i) 3 and 2 are predecessors

$c(2, 1) \times c(1, 1) =$

$\dfrac{2!}{1!1!} \times 1 \qquad = 2$

3 1 1   are successors
$c(3, 1) \times c(2, 2) =$

$\dfrac{!3}{1!2!} \times 1 \qquad = 3$

combine $= \dfrac{6}{60}$

(ii) 3 1 1 are predecessors $= 3$
3 2 are successors $\qquad = 2$ $\dfrac{6}{60}$

therefore 1 is a pivot $\dfrac{6}{60} + \dfrac{6}{60}$ or $\dfrac{12}{60}$ of the time.

$$\dfrac{12}{60} = 20 \text{ percent, or } 6\tfrac{2}{3} \text{ percent for each single vote.}$$

2. Number of times a two-vote bloc is a pivot:

   (a) 3 1 2 3 1 1
             *

   (b) 3 1 1 2 3 1
                *

   Calculate as in 1: $\dfrac{6}{60} + \dfrac{6}{60}$ or $\dfrac{12}{60}$ of the time.

   $\dfrac{12}{60}$ = 20 percent for the two-vote bloc.

3. Number of times a three-vote bloc is a pivot:

   (a) 3 3 2 1 1 1
          *

   (b) 3 2 3 1 1 1
             *

   (c) 3 1 1 3 2 1
                *

   (d) 3 1 3 2 1 1
             *

   (e) 1 1 1 3 2 3
                *

   (f) 2 1 1 3 3 1
                *

   (g) 2 1 1 1 3 3
                   *

   (h) 2 1 3 3 1 1
             *

   Calculate as in 1:

   $\dfrac{4}{60} + \dfrac{2}{60} + \dfrac{6}{60} + \dfrac{6}{60} + \dfrac{2}{60} + \dfrac{6}{60} + \dfrac{4}{60} + \dfrac{6}{60} = \dfrac{36}{60}$ of the time.

   Check: $\dfrac{12}{60} + \dfrac{12}{60} + \dfrac{36}{60} = \dfrac{60}{60}.$ $\dfrac{36}{60} = $ 60 percent, or 30 percent for each three-vote bloc.

Much more spectacular variations from proportionality are easy to construct and are present in the real world (for example, the Security Council of the United Nations). Some cases occur in which some voters have no power, that is, can never be in a pivot position.[11]

---

[11] See John F. Banzhaf, "Weight Voting Doesn't Work: A Mathematical Analysis," *Rutgers Law Review* 19 (1965): 317–340; and Robert G. Dixon, Jr., *Democratic Representation: Reapportionment in Law and Politics* (New York: Oxford University Press, 1968), pp. 537–543.

The Shapley value is particularly useful in analyzing new institutions because of the "all permutations equally likely" assumption. In more localized situations, this assumption can be modified to reflect empirical evidence of different probabilities for coalition formation.

The Basis of Representation

Representation of people in a legislature is usually on a territorial basis; that is, people residing in one geographic area compose a representative unit. While this practice has a good deal of history behind it, it is not the only way, historically, that representation has been achieved. Classes of society can conceivably be represented and were during periods of English history. The House of Lords is perhaps the last lingering shadow of that practice.

Territorial representation was decisively chosen by the framers of the American government, however, and the decision has more than history to recommend it. Writing in the nineteenth century, Bagehot noted that a functional, or interest, representation "would be a church with tenants; it would make its representative the messenger of its mandates and the delegate of its determinations."[12] A deliberate body composed of such inflexible (by definition) messengers has two options. Either nothing is agreed upon, or side payments (trades) must be made on a far larger scale than would be necessary if representation were on a territorial basis. That this is necessarily so may be seen by considering two minimum winning coalitions, one territorial and one functional. To assemble a territorial coalition on a group of issues, some benefit must accrue to each member. The benefits may be in any functional area (a park in one area, better road maintenance in another). To assemble a functional coalition, some benefit must accrue to each function in the coalition (better road maintenance everywhere, more parks everywhere). The result is better described as a division of the spoils than as legitimate legislative trading related to intensities of preferences of citizens.[13]

The foregoing discussion should be distinguished from the dissimilar case

[12] W. Bagehot, *The English Constitution*, 2nd ed. (London: Kegan Paul, Trench, Trübner and Co., 1905), p. 156.

[13] Madison, unlike some of his modern interpreters, was not an advocate of representation of interests. What he said was that interests receive appropriate recognition through territorial representation. See *The Federalist*, Paper no. 10. For an even more thorough anti-interest representation view, see Hamilton's comments in *The Federalist*, Paper no. 35.

of "representation" in an executive or committee situation. In a company, for example, it is quite "correct" and efficient to have representatives of marketing, production, and accounting participate in making a decision. It is equally efficient to have spokesmen for industry, labor, and the consumer advising an executive or administrative agency of government on the most efficacious way of *implementing* a given policy. In the business example, the objective function (profit maximization) is not in dispute. In the executive agency example, policy is presumably already given by the legislative authority for the agency.

None of this is an assertion that "interests" do not dominate many existing legislatures; it is rather a statement that, if they do, representative government is not functioning as it was designed to function, and it is an indication that the failure is probably costly.

Assuming that we do not have to contend with functional representation, we still do have to contend with long-term confusion concerning the two-party system and proportional representation. This confusion essentially starts from the simplistic reasoning that, since 51 percent of the electorate may elect, *therefore* 49 percent of the electorate is unrepresented. If this premise is accepted, the next step is to devise schemes of multiple seats per district to allow a greater proportion of the electorate to be "represented." Lewis Carroll was preoccupied with this problem, as Duncan Black has shown us.[14]

It can be shown, however, that a two-party system has the unique characteristic of being able to represent all the electorate (see appendix C). That is to say that for given issues, it throws up those two positions which, when voted upon, result in the same choice as would have been chosen if *all* the voters were assembled and capable of exploiting vote-trading possibilities. Let me illustrate by the following example.

Suppose three voters and two independent issues as:

		Voter	
Issue	I	II	III
A	Y_2	Y_1	N_1
B	N_1	Y_2	Y_2

[14] Duncan Black, "Lewis Carroll and the Theory of Games," *American Economic Review* 59 (1969): 206–210; and Black, "The Central Argument in Lewis Carroll's 'The Principles of Parliamentary Representation,' " *Papers on Non-Market Decision Making*, vol. 3 (Charlottesville: University of Virginia, 1967).

where Y is a "yes" vote, N is a "no" vote, and the subscripts are the ordinal rankings of the issues by each man.[15] Were these men meeting in an assembly to decide (by majority vote) these two issues, issue A would pass and issue B would fail, even though the nominal outcome would appear to have both issues passing (two Y's on each row). But since it is in the interest of voters I and III to agree to a trade (voter I voting N on issue A in exchange for voter III voting N on issue B), the nominal outcome does not hold. The trade, however, is not stable, since voter II, rather than allow the trade to take place, agrees to give up his vote on issue B to keep voter I from trading. Hence, issue A passes, issue B fails.

Were we to suppose that the three men, not meeting in an assembly, simply vote for one or another candidate, we can see how a two-party system works to achieve the same result. In an election process, voters are faced with a set of mutually exclusive choices. Based on the above matrix, these outcomes can be displayed as follows:

	Alternative Outcomes			
Issue A	P	P	F	F
Issue B	P	F	P	F
Voter I wins	2nd	1st 2nd	none	1st
Voter II wins	1st 2nd	1st	2nd	none
Voter III wins	2nd	none	1st 2nd	1st

where P is pass, F is fail, and the numbers are the ordinal preferences won under each outcome by each voter. (Hence, under $\left[{}^{P}_{P}\right]$ voter I wins his second choice, under $\left[{}^{P}_{F}\right]$ he wins both his first and second choices, etc.)

Knowing the nominal outcome (the typical information available to parties through opinion polls), the candidate of one party may, by testing which issues evoke most response, decide that he can win over $\left[{}^{P}_{P}\right]$ by attracting voters I and III on a $\left[{}^{F}_{F}\right]$ platform, this being the only outcome that dominates $\left[{}^{P}_{P}\right]$. Facing that platform, the second candidate finds that the outcome $\left[{}^{P}_{F}\right]$ can win over $\left[{}^{F}_{F}\right]$ and chooses it. As between the two, he is right. (Were a third candidate to appear, no clear dominant outcome would be possible.)

---

[15] Thus, voter I prefers defeating issue B to winning on issue A; voter II prefers winning on issue A to winning on issue B; voter III prefers defeating issue A to winning on issue B.

Voter III will thus gain neither of his choices, no matter whether he is in the legislature himself or votes in a two-party election. His misery lies not in the fact that he is not represented, but in the fact that there is no way in which he can gain adherents to his positions on these issues.

In the paper cited (see appendix C), I have shown that the congruence between the election results of a properly functioning two-party system and the decision of a legislature composed of the same voters extends at least through all nontrivial permutations of the two-issue/three-voter cases (8 examples) and the three-issue/three-voter cases (432 examples).

The Decision Rule for Legislative Bodies

The long history of simple majority rule for legislative bodies leads one to speculate that its roots go back to primitive days when physical force was the decision principle. However, Douglas Rae and Michael Taylor have recently explicated the logic of simple majority and special majority decision rules.[16] Their contribution is especially noteworthy since economists, following Wicksell, have had nothing to say on this matter other than to agree that unanimity is a sufficient condition for optimality while admitting that it suffers a little in practical application. Rae shows that, if an individual is equally interested in minimizing the summed frequency of rejection of proposals he favors and passage of proposals he is against, then simple "majority rule corresponds uniquely to the minimum summed frequency [of these events] in committees with an odd number of members, and shares the minimum with $n/2$ in committees with an even number of members." This is shown in table 1.

Rae's analysis also allows us to value differently the possibility of losing on a supported proposal and winning on an opposed proposal. In other words, in some cases we may be more interested in keeping "bad" actions by others from occurring than we are in getting an action we approve of passed. An example is amendments to the Constitution. We require two-thirds majorities rather than simple majorities. Why? Because we are minimizing the possibilities of bad actions, that is, actions which are disapproved by even one-third of the voters. In effect, we are assigning double

[16] Douglas Rae, "Decision Rules and Individual Values in Constitutional Choice," *American Political Science Review* 58 (1969): 40–56; Michael Taylor, "Proof of a Theorem on Majority Rule," *Behavioral Science* 14 (1969): 228–231.

Table 1. Expected Summed Frequencies

k^b	n^a 3	4	5	6	7	8	9	10	11	12
1	.43	.47	.48	.49	.50	.50	.50	.50	.50	.50
2	.28	.34	.39	.43	.46	.47	.48	.49	.50	.50
3	.43	.34	.32	.35	.39	.38	.45	.47	.48	.48
4		.47	.39	.35	.34	.36	.39	.42	.44	.46
5			.48	.43	.39	.36	.36	.38	.40	.42
6				.49	.46	.38	.39	.38	.38	.39
7					.50	.47	.45	.42	.40	.39
8						.50	.48	.47	.44	.42
9							.50	.49	.48	.46
10								.50	.50	.48
11									.50	.50
12										.50

Source: Douglas Rae, "Decision Rules and Individual Values in Constitutional Choice," *American Political Science Review* 58 (1969): 40–56.

Note: Italicized probabilities show minimum in each row. Read for values of k in relation to n.

  [a] n = number of members.
  [b] k = number of members necessary to impose a policy on the group.

weight to bad actions when we use a two-thirds majority rule. Rae shows this for $n = 7$ as follows:

k	n 7
1	1.00
2	.91
3	.72
4	.51
5	.45
6	.47
7	.50

Here the expected frequencies are unequally weighted (one for ability to pass legislation, two for ability to defeat legislation). Contrast this column, in which $k = 5$ at the minimum, with the corresponding column in table 1, where $k = 4$ at the minimum. Should we want to attach different weights, we can find the appropriate decision rule by Rae's formula.

Decision-Process Models

One way to look at a decision process is to posit, or assume, a structure (government) and then look at the way in which men and interest groups move, coalesce, or lobby to get what they want. The emphasis is on personality, social groupings, and power. Much insight is gained about the particular circumstances of a case in this way.

Another way is to posit, or assume, that men and groups try to attain their goals (at all times and in all societies) and then look at the way a particular structure, or set of rules, shapes the way that men and groups must act to get what they want. Much insight is gained about the general influence of the institutional arrangements in this way. The decision models we shall be looking at are all of the latter kind. Institutions are easier to change than is human nature.

We have so far been dealing with rather static concepts. They were appropriately static, however, because they pertained to the construction of the basic building blocks of a process. Increasingly, we are gaining insight into the process itself, and it will be useful to examine how various people have done that.

Regression Analysis. Several writers, notably Birdsall, Kramer, and Jackson, have made use of multiple regression analysis to explain voting behavior.[17] While no explicit use will be made of regression analysis in this paper, it is fundamental to many decision models. Jackson's work, an analysis of the voting behavior of U.S. senators both in general and on specific bills, is a good example of what may be learned from such analyses.

Using such independent variables as constituency, party position, majority and minority leadership, and committee position, Jackson was able to explain up to two-thirds of the variation in senatorial voting behavior. Different senators require, of course, different variable specification, but much variation turns out to be regional (particularly among the Democrats). Jackson extended the analysis by examining the residuals from the

[17] W. C. Birdsall, "Public Finance Allocation Decisions and the Preferences of Citizens: Some Theoretical and Empirical Considerations" (Ph.D. diss., Johns Hopkins University, 1963); Gerald Kramer, "Short Term Fluctuation in U.S. Voting Behavior: An Econometric Model," mimeographed (Yale University, 1967); John E. Jackson, "A Statistical Model of United States Senators' Voting Behavior" (Ph.D. diss., Harvard University, 1968).

specified models explaining senators' votes in order to find differences among specific bills. Since logrolling (vote trading) was not a part of the regression equations, this technique can reveal whether vote trading might have played a role on some bills but not on others. In particular, if the regression equation explains a great deal of a senator's voting behavior on bills known to be of high interest to him, while explaining little of his voting behavior on bills known to be of less interest to him, one can surmise that he may be trading off his votes on the latter cases. Jackson found four bills in the Eighty-seventh Congress with relatively large unexplained variance, all of which were "bills which evoke an image of special interest, vote trading, and bargaining." They were the Sugar Act amendments of 1962, elimination of the West Coast Maritime Subsidy Preference, establishment of a National Wilderness Preservation, and the Public Works bill.

While cautioning that there may be other explanations for these results, Jackson concludes that vote trading may have accounted for at least a part of the poor performance of the regression equations in these cases. There is the further possibility that some of the vote trading is masked by the specification of the models themselves.

Regression analysis and its more complicated relatives (factor analysis and principal components analysis) will obviously play an increasing role in analyzing past actions and in giving some insight into possible future actions.

The Significance of the Mean. Since Black[18] demonstrated that the median position can win over any other position under majority vote and single-peaked preference conditions (analogous to Hotelling's location theorem for stores along a single street), there have been explorations of the significance of the median and mean positions for political strategy. In a series of articles, Davis, Hinich, and Ordeshook expanded this theme into multi-dimensional space.[19] The work proceeded along two lines. First, following

[18] Black, *Theory of Committees*, p.18.

[19] Otto Davis and Melvin Hinich, "A Mathematical Model of Policy Formation in a Democratic Society," in *Mathematical Applications in Political Science*, vol. 2, ed. Joseph Bernd (Dallas: SMU Press, 1966); Davis and Hinich, "Some Results Related to a Mathematical Model of Policy Formulation in a Democratic Society," in *Mathematical Applications in Political Science*, vol. 3, ed. Joseph Bernd (Charlottesville: University Press of Virginia, 1967); Davis and Hinich, "On the Power and Importance of the Mean Preference in a Mathematical Model of Democratic Choice," *Public Choice* 5 (1968): 59–72; Melvin Hinich and Peter Ordeshook, "Social Welfare and Electoral Competition in Democratic Societies" (Paper presented at Public Choice Society, Chicago, 1970).

Arrow, they expanded the set of individual utility functions (beyond single-dimensional ones), for which necessary and sufficient conditions for transitive, social preference ordering could be determined. This work will not be discussed here. Second, they showed that in a variety of multivariate distributions of preference, that position (a multivariate mean) which minimizes the Euclidean distance from all individual positions is a social welfare maximization (weighting all individuals equally). Where median and mean diverge, therefore, a candidate can achieve a dominant position at the median on all issues, while a social welfare "solution" would be at the mean.

While the Davis, Hinich, and Ordeshook theorems are undoubtedly correct, the assumption of independence between issues and intensities of interest suppresses an important feature of politics, as Jackson's critique[20] of the Davis and Hinich works makes clear.

It may be helpful to explain this point in words and with the aid of a diagram.[21] Imagine, for purposes of exposition, a two-dimensional situation (two issues) and only two voters. Suppose that one voter feels very strongly about one of the issues but not about the other, while the other voter has a similar set of preferences but about the opposite issues. Thus in figure 1 the preferred or ideal points of the two voters are indicated by x_1 and x_2 respectively. The above assumption means that the indifference curves (with increasing loss indicated by curves further from the ideal points) of one voter are elongated in the east-west directions while the other's indifference surfaces are elongated along the north-south dimensions. The mean x_1 and x_2 falls along the dotted line connecting the two points and is indicated by the point μ. Clearly, however, point μ is inferior from the viewpoints of both of the voters to some other point, such as the point y, which allows each to have something closer to the ideal amount of the more desired item at the expense of the unimportant one. Hence, in this example the mean μ is neither a dominant point nor a welfare maximum (loss minimum). This point obviously is important whenever there is an "opposition" of the kind shown in this example.

Since neither preference positions nor preference intensities of voters are known in most practical cases, perhaps we should not concern ourselves with the problem at all. In fact, of course, most politicians realize that dis-

[20] John E. Jackson, "Intensities, Preferences, and Electoral Politics," Working Paper 705–772 (Washington, D.C.: The Urban Institute, March 1972).

[21] I am indebted to Otto Davis for this illustration, which I have taken from his written comments on my paper.

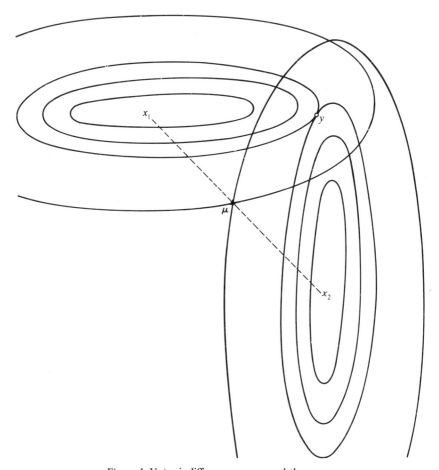

Figure 1. Voter indifference curves and the mean.

covering such information is crucial to being elected. Moreover, "defining" issues so as to change voters' preferences and intensities in one's favor is at the heart of the matter.

This point may better be grasped by illustration. Pollution may be regarded as a current problem of intense interest, but the way in which issues are formed about the problem will greatly affect the distribution of intensities of interest. A proposal to institute a system of effluent charges will evoke one set of intensities; a proposal to subsidize treatment plants will evoke another; a proposal combining both approaches will evoke a third.

Varying the amounts and incidence of either charges or subsidies will affect intensities further. These are not the same as the position means brought forth by different ways of forming the issue around the pollution problem. The man or group who is successful in forming the issue determines the relative (ordinal) ranking that the problem will take, not only in his preference orderings *but also in everyone else's.* This is why the committees that determine which bills reach the floor are so powerful and why the process of issue formation among the electorate is so important. If I, as a candidate, can define an issue so that it divides your supporters while uniting and adding to mine, then making such definitions has high payoff. Examples are numerous. One of recent years was the deliberate linking of the church-state separation issue with federal aid-to-school proposals. The linkage split off some support for the school aid proposal. The point here, it must be emphasized, is that the bill's position on school aid was no farther from the former supporters' position than it had been, but intensity of interest in supporting the bill had fallen because the religious issue had been linked to it. Since intensities of interest around the general church-state problem and around the school aid problem were not distributed independently of the issue position on each, finding a multivariate mean (loss function) on the two-issue space serves, not to specify a preferred candidate position, but to respecify the issue. Candidate positions will be taken in response to intensities of interest for and against each such redefined issue. (See appendix C for further discussion of issue formation.)

The Role of the Party in Issue Formation. The two-party system in twentieth-century America has played less than its appropriate role in issue formation, even though the utility theory of representative government depends upon a two-party system that is able to formulate the issues appropriately. American government depends much more on parties than we suppose. We neglect parties at the peril of having to contend increasingly with intransigent, nonnegotiable demands made by one or another faction within our society. A government can "cope" with demands, but such actions should not be confused with the process of self-government under any definition. In Burke's words, "Men of intemperate minds cannot be free. Their passions forge their fetters."[22]

[22] Edmund Burke, *A Letter to a Member of the National Assembly in Answer to Some Objections to his Book on French Affairs* (1771).

Issue formation will reflect individual utilities if there is a functioning party structure. No other group or organization in society has this task. The press has no warrant to do more than raise cries of alarm and argue points of view. Special interest groupings have no interest beyond their own. Neither has any responsibility for decisions and cannot be made responsible. The executive branch of any government cannot be expected to form issues correctly, because it is controlled by one party at a time.

While all of these groups and many others contribute to the formation of issues, only the party system that runs candidates and, when elected, is identified as responsible for programs has the task and the incentive for defining the area of battle over time. Parties can, moreover, operate in time horizons longer than one term, unlike individual candidates with no strong party backing. In the United States parties have the role held in Bagehot's England by the House of Commons,[23] and perform the following functions:

1. The expressive function—reflecting what is on the public mind.
2. The informing function—informing us (and the government) of what we would not otherwise hear (since there are many publics and we know only a few of them at first hand).
3. The teaching function—restating the crude and formless worries, prejudices, and offhand opinions of the public into reasonable, defensible proposals.
4. Coalescing function—forming protocoalitions around each proposal.
5. The elective function—presenting final positions from which individual choices must be made by the voter.

This deliberately chosen "old-fashioned" statement of moderate men engaged in moderate action stands in sharp conflict with the politics of the 1960s, a time of growing estrangement from traditional party politics.

It is worth a moment to dismiss the claim (frequently made by some professionals) that the role of the professional expert preparing unbiased plans is a surrogate for party positions. This modern variant of the "there is no Republican way to collect the garbage" fallacy can best be illustrated by the failure of the Brandywine Plan,[24] in which it was found (again) that planning plus public relations is not the same thing as self-government.

---

[23] Bagehot, *The English Constitution*, pp. 132–175 *passim*.

[24] See Peter Thompson, "Brandywine Basin: Defeat of an Almost Perfect Plan," *Science* 163 (March 1969): 1180–82.

The point is a simple one. If we do not use the party system to ensure that moderate men will take moderate action, what other recourse do we have?

Minority Interests and the Dominant Majority.   Closely allied to the role of parties in forming the issues is the role of general-purpose legislatures as a device that protects against a dominant majority. In a recent paper (see appendix B), I have shown how Madison's defense of representative government as a way to avoid "the tyranny of the majority" is justified in general-purpose legislatures far more than in special-purpose commissions. Essentially, with more independent issues to consider it is less likely that any one coalition will agree on all issues. If they do not, then vote trading can allow minority interests to achieve (or block) legislation of greatest concern to them. The two-party system does not interfere with this vote trading, since by definition a two-party system is composed of nondoctrinaire, rather loose associations of many interests. (With doctrinaire parties, as in proportional representation, such trades are greatly inhibited.)

As the number of independent issues is reduced, trading is inhibited and thus many commissions may show records of all issues being decided by the same majority for long periods. Minority "representation," in these cases, is a sham.

The same frustrated minority issue can easily arise in a city in which the council is elected at large rather than by wards. A minority may not be able to get into the council at all under such circumstances. If, in addition, the city functions on a "nonpartisan" basis, a minority may have no way to bring pressure on officials other than by violence.

Variation on the Utility-Maximizing Theme.   So far all of the discussions on the decision process have implicitly assumed that there are utility-maximizing individuals and groups. Standing somewhere between this approach, common to economics, and the full-scale bargaining models applied to international relations (in which threats and outright irrationality may play a larger role) are some attempts to make explicit what a political philosopher would recognize as a theory of the state. Kenneth Boulding has become increasingly concerned with distinguishing selfishness and selflessness.[25] While at first blush his malevolence-benevolence distinction seems to be only a fruitless inquiry into the *motivation* of behavior and thus

[25] Boulding, "The Network of Interdependence."

Table 2. Payoffs of Collective Decision Processes

Distribution			Sum of payoffs before decision	Increment in sum of actors' payoff due to collective decision		
				Social welfare	Individual bargaining	Competitive bargaining
0.9	0.1	0.0	496	332	52	−140
0.6	0.3	0.1	682	10	62	−88
0.5	0.3	0.2	688	10	88	37

Distribution			Inequality before collective decision	Decreases in inequality		
				Social welfare	Individualistic	Competitive
0.9	0.1	0.0	443	250	288	363
0.6	0.3	0.1	252	50	11	83
0.5	0.3	0.2	199	56	−59	46

Effects on Inequality (average deviations)

Source: Rufus R. Browning, "Quality of Collective Decisions: Some Theory and Computer Simulations," mimeographed (Michigan State University, 1969).

falls under the criticism that the utility maximizer's motivation is not at issue, Boulding's analysis is not to be so dismissed. His concern is with showing that the consequences for social choice are different depending on how benevolence or malevolence influences individual utility functions.

This general area has been explored by Rufus Browning.[26] By simulation, he examines the consequences of positing different rules by which collective decisions are made—that is, "selfish" utility by each individual, maximizing overall welfare (summing cardinal utilities), and competitive utility (maximizing the difference between one's own and others' utility). Different probabilities of being in the minority are assigned to each actor in each process. Strikingly different results are obtained. For example, table 2 relates the effects on total payoffs (summed over all actors) of three collective decision processes for three utility distributions. Consequences for individuals are also quite different under the three choice processes.

The relevance of this work is in its ability to guide the design of a flexible decision structure. For example, if an economically efficient charge is com-

[26] Rufus R. Browning, "Quality of Collective Decisions: Some Theory and Computer Simulations," mimeographed (Michigan State University, 1969).

bined with a receipts distribution on a per capita basis, there will be a different response than there would be if the receipts were not so distributed. The receipts distribution formula gives a "social welfare" or maximum-sum rationality. An example of the perverse workings of this approach is found in the "per capita" component of highway fund disbursements.

Another possibility along the lines Browning suggests is consideration of using a competitive bargaining scheme (maximizing difference between an actor's and others' utility) regarding the amount of effluent any one firm can discharge into a watercourse at a particular point. Converting the effluent to some standard unit, a zero-sum game is involved (assuming some water quality standard) that the firms may play given an appropriate set of rules. The incentive would be somewhat different from that which results from allowing bids for the available assimilative capacity (which implies no interest in capacity beyond one's own need). Under Browning's "competitive" rules, each firm would have an equal interest in denying the capacity to others and in garnering the capacity for itself. Hence, in some situations a firm's interest would be precisely consonant with that of a conservation group! Likewise, the incentives in a cooperative (as is the German *Genossenschaften* system of river management in the Ruhr) are different from the individualistic approach.

Some Specific Critiques

Using the theory and tools of analysis from the preceding section, I shall now examine some specific examples of institutions that have as their primary purpose the making of environmental choices.

The Potomac River Basin Commission

Governmental jurisdictions at local, state, and federal levels are now in the process of designing an institution to make environmental choices regarding the uses of Potomac River Basin water. The proposed federal-interstate compact would establish a river basin commission, composed of the governors (or their representatives) of West Virginia, Virginia, Pennsylvania, and Maryland as well as appointed representatives for the federal government and the District of Columbia.[27] The compact is patterned

[27] For a general survey, see Alvin C. Watson, "A Proposed Partnership Compact for our Nation's Rivers," *Journal of Soil and Water Conservation* 24, no. 3 (1969): 89–93.

Table 3. Cost of Alternative Systems for Improving Dissolved
Oxygen in the Potomac Estuary

(million dollars)

System alternative	Cost	Cost to area
A	22	14
C	29	29
D	38	26
H	115	0

Source: Robert K. Davis, *The Range of Choice in Water Management: A Study of Dissolved Oxygen in the Potomac Estuary* (Baltimore: The Johns Hopkins Press for Resources for the Future, 1968), table 24.

closely after the Delaware River Basin Compact despite the substantial differences between the two basins and their problems. The commission would, in Alvin Watson's words, ". . . by its very existence provide a mechanism to weigh alternative values associated with different plans or components of plans when conflicts in values, uses, or components arise."[28] There is no doubt that this is the case. The question is, is it the appropriate mechanism?

The case for coordinated or unified management of the river is compelling. Kneese and Bower have investigated the economic efficiency gains from systemwide management and Davis has applied these principles to the Potomac.[29] Were efficiency gains the only issue, river basins could well be turned over to any efficient organizational form—a public corporation, a franchised monopoly, or a cooperative.

The actual situation is, of course, reversed. We have a long, involved, and *in-place* system for dealing with equities, and government investment has traditionally been at least as concerned with distribution of the public investment portfolio as it has with economic efficiency, which is a newcomer to the public calculus and finds some difficulty in dislodging older interests and procedures. In table 3 Davis illustrates a part of the problem in the context of the alternative technological possibilities for achieving a given level

[28] Ibid., p. 92.
[29] Allen V. Kneese and Blair T. Bower, *Managing Water Quality: Economics, Technology, Institutions* (Baltimore: The Johns Hopkins Press for Resources for the Future, 1968); Robert K. Davis, *The Range of Choice in Water Management: A Study of Dissolved Oxygen in the Potomac Estuary* (Baltimore: The Johns Hopkins Press for Resources for the Future, 1968).

of dissolved oxygen in the Potomac. Such a distribution of costs might suggest that system H would be an overwhelming favorite of area residents. Were the costs listed the only ones incurred, this would no doubt be the case. However, system H, essentially a series of dams and reservoirs, would have a widely differential effect on residents. Some would be flooded out, others overrun by tourists and vacationers, and still others assured of a more reliable water supply. There was, in fact, such substantial opposition to the plan that the incidence of its effects is being radically shifted.

It is worth noting, in passing, that the agency responsible—through formal public hearing—for generating, recording, and "assessing" public reaction to plan H was the Corps of Engineers of the Department of the Army. While such tasks may teach humility to the Corps (as it would to any body of men on the receiving end), the rationale for putting such a requirement on the Corps can be justified only by the absence of any representative government to which citizens might turn. Substituting an appointive board for the Corps does not lessen the sham that consists of using the form of representative government to deny the substance of it. Officials are either directly responsible to the electorate over whose hearings they preside or they are not. Appointed officials and executive branch personnel are not.

Distributional effects of public policy and equity considerations are important aspects of river basin policy. Such considerations are not necessarily consonant with economic efficiency considerations. To these three elements we should add a fourth—reallocation of resources for some public policy objective. To keep the distinction between this element and redistribution of income, think of the difference between the objectives of the space program and the objectives of the Appalachia program.

Thus the four major objectives for river basin investment are:

1. Economic efficiency or efficient allocation of resources—where the criterion is to put investment where the aggregate payoff is greatest in economic terms *or* where the aggregate economic cost is least if some other objective function is externally imposed.
2. Equity—redressing the balance on the incidence of benefits and costs on programs taken in pursuit of economic efficiency.
3. Income distribution—where the criterion is some specified redistribution of income through programs having a differential impact of costs and benefits.

4. Reallocation of resources—where resources are deliberately forced into noneconomic patterns for a specific public policy purpose; to increase public recreation facilities, for example, which, while it has a general welfare orientation, is focused on the provision of the good rather than on any particular redistribution effects for specific groups of people.

While there is some substantial, operational overlapping of objectives 3 and 4, the conceptual distinction may be useful. Only the first of the four is a management issue; the other three are clearly social choice questions where conflicts in values will certainly occur. Consequently, the interstate compact, in its present form, is suspect as a political mechanism for making social choices.

Some 4 million people live in the Potomac River Basin. Disregarding, for the moment, the 800,000 District of Columbia residents who have no political representation anywhere, the other basin residents do not dominate any state government of the area. Unlike some other river basins, where nearly everyone in the basin states is in the basin, Potomac Basin residents make up, at most, 27 percent of a single state's population and 36 percent of its area, as shown in table 4.

The basin residents do not nominally control either house of any of the legislatures. When one adds the fact that the voters of each state who live in the basin are by no means unanimous on what they want in terms of water choices for the Potomac and thus will not function as a bloc, it is obvious that the state legislatures do not have a clear a priori claim as the appropriate bodies to resolve the issues.

Even less appropriate is the typical interstate agreement among sovereigns. While the proposed compact would give each state one vote, along

Table 4. State Areas vs. State Populations in the Potomac Basin

State	Percent of state area in basin	Percent of state population in basin
Maryland	36	27
Pennsylvania	4	2
Virginia	14	23
West Virginia	15	8

Source: Potomac River Basin Advisory Committee.

Table 5. Potomac Basin Population

Within	Total	Percent of total population in basin
District of Columbia	811,000	22.6
Maryland	1,339,000	37.4
Pennsylvania	185,000	5.2
Virginia	1,103,000	30.7
West Virginia	148,000	4.1

Source: Potomac River Basin Advisory Committee, 1969 estimates of population.

with one vote each for the District of Columbia and the federal representative, the population figures (basinwide) show a far different proportionality (see table 5).

The fact that the river basin is not coterminous with state boundaries (and nowhere dominates any of the states) lends weight to the thought that the interstate compact in its usual form is not wholly appropriate to the case. It is, of course, necessary that state power be brought into the picture. The compact could be made to safeguard the interests of basin residents through a requirement that the states delegate power to basin residents, who then would elect the members of the commission. They would vote on the basis of equal-population districts (congressional districts furnish a suitable, ready-made structure closely coinciding with basin boundaries). In addition to the one federally appointed member, representation on the commission would then be on the basis of about 400,000 people per vote for the other ten members. Maryland (Cong. Dist. 5, 6, and 8) and Virginia (Cong. Dist. 7, 8, and 10) would have three votes each; Pennsylvania (Cong. Dist. 12) and West Virginia (Cong. Dist. 2), one each; and the District of Columbia, two votes.

There is no legal reason why the states could not negotiate a compact that would include such a representative pattern, nor is there any legal reason barring the Congress from approving it. Were this to be done, the compact would undoubtedly also provide that any state funding would come from assessment of basin residents only, rather than from general state revenues. Such a provision would not only be essential for equity but fundamental to fiscal responsibility. A well-known defect in federal grant-in-aid programs is that the money is usually available only for specific technical remedies, thereby reducing the incentives of states and localities to use any other remedies, even though they might be more efficient. Thus federal monies are available for dams but not for reaeration treatment. What state would give serious consideration to the latter under such circumstances? Likewise, if basin residents receive funds from their state's general revenues, their choices will be biased in the direction of those features funded by the state. Moreover, to the extent that costs of any sort can be transferred to the general state or federal taxpayer, these costs will not be fully considered by basin residents.

An interstate compact setting up a basin representative commission would not, of course, solve problems created by grants-in-aid tied to particular solutions. Moreover, a basin-based commission could create some administrative problems within each state. Which of the three Virginia representatives on the proposed commission would be responsible to deal with the governor in Richmond? How is basin water policy to be "coordinated" with the state official responsible for water matters?

Answers to these questions lie in the interstate compact itself. If a representative basin commission were established, it would have to have a clear delegation of state authority regarding water policy and a sharing of state authority in the recreation area. In the water management field, the state would cease to act administratively within the basin (having delegated its authority to do so to the basin commission). In the recreation area, the role of the state (like that of the federal government) would be limited to funding facilities over and above those desired by the commission (to take account of nonbasin needs). This latter issue, while providing a planning problem, is no different in principle from any federal park in a state or a state park in a particular county. Such administrative problems are minor compared with the larger problem created by the present interstate compact proposal, which ignores real basin representation.

Attempts to provide for basin representation by advisory committees dealing with recreation, land-use planning, and other special interests do not serve as surrogates for representative government in weighing values. Any citizen has many interests. He may be an avid fisherman *and* a pinchpenny taxpayer. He may be a land speculator *and* a supporter of free recreation. If citizens are represented only by their special-interest groupings (either by committees or by associations and clubs), the resulting pressures do not necessarily converge toward any real public choice. It is when each citizen must resolve, within his own mind, how he feels about all

Table 6. Nominal Voting Power, by State, under Different Decision Rules

Area	Number of votes	Shapley values in percent		
		Votes needed to pass = 6	Votes needed to pass = 7	Votes needed to pass = 8
Maryland	3[a]	30.00	28.33	28.33
Virginia	3[a]	30.00	28.33	28.33
District of Columbia	2	20.00	18.33	13.33
Pennsylvania	1	6.66	8.33	10.00
West Virginia	1	6.66	8.33	10.00
Federal representative	1	6.66	8.33	10.00
Total voting power	11	100.00	100.00	100.00

Note: Totals do not add due to rounding.
[a] Assuming a bloc vote in order to test dominance possibilities.

aspects of the river basin that choices begin to converge, extreme positions fall, and public choice becomes realistic. But the citizen, speaking as a whole man, can do so only if he can vote. Since he cannot vote on all the technical issues (not because they are technical but because there is no way to set up an efficient vote-trading mechanism), he should vote on his choice for a representative. If representatives are to throw up the correct choices for him, then the two political parties should be sponsoring the candidates.[30] It is sometimes argued that voters respond very poorly to special districts, both by allowing political machines to capture all nominations and by ignoring the elections in droves. Such arguments cut both ways, however. Many special districts are poorly conceived and deal with very narrow issues that give the voter no clue as to their relevance to his interests. If a basin commission is given broad authority to act, the many citizen interests now frustrated in the "public hearing" charade will not be reluctant to make very live election issues out of the real alternatives and candidates offering themselves for election.

If we assume the existence of such an elected commission (10 elected officials and an appointed federal representative), then we can make some judgments about the decision rules under which the commission might operate. First, using Shapley values, we can test the bloc power by states (see table 6). The interesting feature revealed here is that attempts to help

[30] It should go without saying that technical analysis must underlie these party positions. To argue that political choice processes are paramount is not to argue that the choices must be based on ignorance of the consequences of alternative actions.

Table 7. Upstream-Downstream Nominal Voting Power
under Different Decision Rules

| | Shapley values in percent | | |
Number and size of blocs	Votes needed to pass = 6	Votes needed to pass = 7	Votes needed to pass = 8
Case I			
1 five-vote bloc	50.00	58.33	50.00
1 four-vote bloc	16.67	25.00	50.00
2 one-vote blocs	16.67	8.33	0.00
Case II			
2 five-vote blocs	33.00	50.00	50.00
1 one-vote bloc	33.00	0.00	0.00
Case III			
2 four-vote blocs	30.00	35.00	50.00
3 three-vote blocs	13.00	10.00	0.00

Note: The power listed in the "one-vote" bloc cases is for *each* such vote.

"one-vote blocs" by increasing the majority required for approving decisions does do so, but at the primary expense of the District voting power
rather than of that of the three-vote blocs.

A more realistic Shapley value relates to upstream-downstream differences. Table 7 makes three separate assumptions about how the upstream-
downstream coalitions would shape up and then tests the voting power
under each. Here the voting power values fluctuate sharply; some changes
in the decision rule cause dramatic shifts in relative power and others have
no effect at all. Other bloc assumptions might be considered in the process
of deciding upon the decision rule which has greatest general acceptance
(i.e., can pass all relevant legislative bodies).

In the case of the Potomac, the issue of control of land along the river
has for several years been a major divisive problem. The work of Rae[31]
concerning special majorities may be useful here. Assuming that land-use
issues will not become a bloc issue (i.e., that each representative's interest
will diverge in some important respects from that of every other representative), then we might give the commission power on land use but require a
larger (say two-thirds) majority for the enactment of any restriction on land
use. Since the two-thirds rule gives double weight to blocking unwanted

[31] Rae, "Decision Rules and Individual Values."

actions, those opposed to the commission's having authority to control land use would have far less to fear.

Alternatively, a less than simple majority could be made sufficient to approve certain uses. Thus, being able to affect in a positive way what is done could also bring agreement among those who at present are reluctant to approve the compact.

There is some merit in including land-use powers (and indeed other environmental issues) in the proposed commission, since a greater variety of independent issues *about which all parties have an interest* is conducive to greater trading potential. It is also the only defense against the problem of the dominant majority. A counsel of perfection would maintain that perfectly functioning general legislatures could make Potomac Basin choices, appropriately coordinated with other legislatures, without needing the compact at all. Lest we fall into that error, we should remember that the necessity to devise any mechanism for making choices in a river basin is, to some degree, the recognition of failure at the state and national legislative (not executive)[32] level. What a properly shaped elective Potomac River Basin Commission could do is to remove the currency of Potomac water projects from the halls of Congress and put it into circulation as the legal tender of an institution appropriate to its reach.

The San Francisco Bay Commission[33]

The social choice problems confronted by persons living around a bay are different from those confronted by persons living in a river basin. Taking the three utility classifications of Browning,[34] for example, we may distinguish the two cases. The classic river problem is the upstream-downstream conflict. The utility framework of the upstream users is most like Browning's "selfish" or individualistic utility; the upstream user is concerned with maximizing his own satisfactions. The downstream user, however, is closer to Browning's "competitive" utility maximizer, one who is forced to maxi-

---

[32] The use of the state governor as the representative of the state on interstate matters is essentially an executive idea. Under no utility theory can this be associated with the aggregation of individual preferences into social choices. It is doubtless political, since the governor is an elected official, but the politics are far removed from the issues at hand.

[33] I am indebted to Joseph E. Bodovitz, executive director of the Bay Commission, and his staff for help in preparing this section and correcting some errors of fact. Mr. Bodovitz is not, of course, responsible for my interpretations or conclusions.

[34] Browning, "Quality of Collective Decisions."

mize the difference between his utility and that of the upstream user. I do not make the distinction to deprecate one side or the other, but to show what roles each is allowed, or forced, to play by the lack of a riverwide platform in which policy may be set.

If we were to imagine that each river basin user was in a compulsory cooperative regarding the uses of the water, we could see that an additional utility concept—overall maximization—would be added to the picture. This is because increasing the total uses of the river will enable both upstream and downstream users to increase their utilities (using their own principles of maximization). River basin commissions have as their underlying motivation the same addition to the utility calculations.

The users of a bay quickly run to the limits of selfish or individualistic utility maximization and become, even without formal machinery, aware that everyone plays a zero-sum game on use and that everyone could be better off (a bigger pie) by cooperating. The interdependence of utilities is perceived more easily, and this is particularly true when the bay is a great scenic resource.

San Francisco Bay is such a scenic resource that it is not surprising that there have been increasingly strict rules regarding its use, despite the large number of governmental jurisdictions and levels of government involved.

The San Francisco Bay Conservation and Development Commission (BCDC) was created in 1965 by the California legislature (the McAteer-Petris Act) and was made a permanent agency by the 1969 legislature. The BCDC has three major responsibilities:

1. To regulate (by issuance or denial of permits) all filling and dredging in San Francisco Bay in accordance with law and the BCDC Bay Plan.

2. To have limited jurisdiction over substantial developments within a 100-foot strip inland from the bay. Within this strip, the commission's responsibility is twofold: to require public access to the bay to the maximum extent possible consistent with the nature of new shoreline developments, and to ensure that existing shoreline property suitable for such high-priority purposes as ports, water-related industry, and water-related recreation, is reserved for these purposes, thus minimizing pressures to fill the bay.

3. To have limited jurisdiction over any proposed filling of salt ponds or managed wetlands (areas diked off from the bay and used for salt produc-

tion, duck-hunting preserves, or similar purposes). These areas, though not subject to the tides of the bay, provide wildlife habitat and water surface important to the climate of the bay area. If filling of these areas is proposed, the commission is to encourage dedication or public purchase to retain the water area. If development is authorized, the commission is to ensure that such development provides public access to the bay and retains the maximum amount of water surface consistent with the nature of the development.[35]

These responsibilities evoke the range of social choices under consideration in the area. The objectives, according to the bay plan, are to maintain and enhance the bay as a "magnificent body of water that helps sustain the economy of the western United States, provides great opportunities for recreation, moderates the climate, combats air pollution, nourishes fish and wildfowl, affords scenic enjoyment, and in countless other ways helps to enrich man's life."[36] It will be noted that while water quality is of concern to the commission, it does not have legal responsibility for regulation of waste discharges into the bay.

Membership on the Bay Commission reflects a complex pattern of local, state, and federal concerns. The twenty-seven members are appointed as follows:

Federal level executive officials
1. One member by the Division Engineer, U.S. Army Engineers, South Pacific Division, from his staff (does not vote on permits).
2. One member by the U.S. Secretary of Health, Education, and Welfare, from his staff (does not vote on permits).

State level executive officials
3. One member by the Secretary of Business and Transportation, from his staff.
4. One member by the Director of Finance, from his staff.
5. One member by the Secretary of Resources, from his staff.
6. One member by the State Lands Commission, from its staff.

[35] Taken from various publications of the San Francisco Bay Conservation and Development Commission.
[36] San Francisco Bay Conservation and Development Commission, *San Francisco Bay: What Will It Be Like in 50 Years?* (January 1969).

7. One member by the San Francisco Bay Regional Water Quality Control Board, who shall be a member of such board.

Local level executive officials

8. Nine county representatives consisting of one member of the board of supervisors representative of each of the nine San Francisco Bay area counties, appointed by the board of supervisors in each county. Each county representative must be a supervisor representing a supervisional district that includes within its boundaries lands lying within San Francisco Bay.

9. Four city representatives appointed by the Association of Bay Area Governments from among the residents of the bayside cities in each of the following areas:
   (a) North Bay—Marin County, Sonoma, Napa, and Solano.
   (b) East Bay—Contra Costa County (west of Pittsburg) and Alameda County north of the southern boundary of Hayward.
   (c) South Bay—Alameda County south of the southern boundary of Hayward, Santa Clara County and San Mateo County south of the northern boundary of Redwood City.
   (d) West Bay—San Mateo County north of the northern boundary of Redwood City and the city and county of San Francisco.
   (Each city representative must be an elected city official.)

"Public" members

10. Seven representatives of the public, who shall be residents of the San Francisco Bay area and whose appointments shall be subject to confirmation by the Senate. Five of such representatives shall be appointed by the Governor, one by the Committee on Rules of the Senate and one by the Speaker of the Assembly.[37]

At first blush, the membership may remind one that fascism (or state syndicalism), in the nonpejorative sense, is not dead but is alive and well in California. Such a body could even make the trains run on time.

This judgment may be modified by a closer examination. There is an absolute majority of elected officials on the commission, though their election is not primarily, or even importantly, predicated on service on the

[37] Government Code of California, Title 7.2, chap. 3, 66620.

commission. There is also an absolute majority of local government members, although it is rare that all local members' interests would coincide.

A more compelling reason for approval of the commission membership is that it is serving an executive function. Basic policy for development and conservation of the bay is spelled out in the McAteer-Petris Act by the state legislature. While set in broad terms, the policy does specify priorities on uses of the bay shoreline, set guidelines in some detail, and specifically endorse and adopt the bay plan. Thus the representation of executive agencies and, in effect, of interest groups can be considered a legitimate device to ease implementation of basic state policy. It would be far different, and less desirable, to set up such a body of men *and* give them authority to write their own basic policy or, worse yet, to allow them to operate without one. As it is, the statutory policy may be amended by the state legislature and appeals to the law can act as a restraint on commission decisions. In effect, the major social choices about the bay are made in the state legislature and the commission is an implementing or executive agency.

The commission is nonetheless interesting on that account, as an examination of its minutes will reveal. With twenty-five members allowed to vote on permit applications (thirteen votes needed to grant a permit), the voting records do not show cases of local officials ranged against other members. As one would expect, they show instead voting patterns based more on geographical and interest lines. Looking at a number of permit cases over the last three years in which the vote margin was only one (therefore a minimum winning coalition existed) revealed that: (*a*) no dominant majority problem exists, i.e., the members of the minimum winning coalition change; (*b*) there is a small pro-permit bloc; (*c*) there is a small anti-permit bloc; (*d*) there are two larger blocs that normally vote together and in opposition to each other.

Tables 8 and 9 do not reflect these blocs as such, but rather explore a larger range of Shapley values for blocs operating in this system.

Contrast the bloc voting power when some voting members are absent with that when all are present, as shown in table 8. If nothing else, such a contrast should increase attendance at commission meetings.

Using some alternative bloc sizes, and assuming full attendance, table 9 explores some of the variations in voting power that are possible. As shown in table 9, adding one member to a bloc of four actually may double the bloc's power, while adding a sixth member does nothing.

Table 8. San Francisco Bay Commission—Bloc Voting Power (13 needed to pass)

	19 present		25 present	
Bloc size	Number of blocs	Voting power (%)[a]	Number of blocs	Voting power (%)
6	1	30	1	27
5	1	22	1	20
4	1	22	1	15
3	1	22	1	10
1	1	5	7	4 (each)

[a] Does not add up to 100% due to rounding.

One feature of the picture inhibits the whole process of vote trading and adds to the feeling that the commission is essentially an executive (non-policy) agency. Since permit applications are not decided on simultaneously, there are continual and nondirected changes in the issue space. Since

Table 9. San Francisco Bay Commission—Possible Variations in Voting Power

	Bloc size	Number of blocs	Voting power (%)
I	11	1	48
	9	1	14
	1	5	7.6 each
II	6	1	25
	4	3	16
	3	1	13
	1	4	3
III	9	1	50
	6	1	17
	4	2	17
	1	2	0
IV	9	1	50
	5	1	17
	4	2	17
	1	3	0
V	8	1	43
	6	1	20
	5	1	20
	4	1	10
	1	2	3

there is judgment about each permit, there is also a record. Quite clearly, consistency has a necessary part to play here unless the courts are to get many of the cases on appeal.

The record of the last three years supports the view that consistency (the executive virtue) has been achieved. Court appeals have not been numerous. Some cases involving many interests have been drawn out over time until a compromise acceptable to many was worked out. The Oakland Airport permit is a case in point.

Having reinforced the case for designating the commission as an executive agency, one may properly ask why fourteen elected officials are on it and whether or not the state legislature was the appropriate policy body to define policy for this "land management" agency for San Francisco Bay. Knowing nothing about the practical politics of the case, I would still venture the opinion that, in the absence of any policy mechanism at the metropolitan level, the state legislature was the only legitimate device that could have been used. Since the legislature endorsed the Bay Plan as developed by the commission, policy was at least initiated locally and the presence of fourteen elected officials on the commission is an indication that policy may well be changed incrementally by commission decision in the future.

If that is true, the commission's structure and operations should be seen as a response to the present realities of the California situation. In particular, the successes and failures of the commission will be sui generis, and the commission form and structure should not be considered a general solution that other areas could copy with profit.

The Penjerdel Air Pollution Case

Making environmental choices about air quality in particular places is even more difficult than making choices about water. Not only are many governmental jurisdictions involved, as in water problems, but also the airshed, unlike the watershed, is not well defined.

Nevertheless, the growing problem of air pollution in urbanized areas makes it essential that the problem become subject to governance by someone, and federal, state, and local governments are beginning to try to cope with the peculiarities of the problem. Unfortunately, the standard formula of the interstate compact among sovereign states is the primary mechanism now being considered. The Air Quality Act of 1967, authorizing the Department of Health, Education, and Welfare (HEW) to establish air quality control regions, also looks to the establishment of traditional interstate compacts to provide the mechanism for control.

In the area that encompasses Philadelphia, Camden, Trenton, and Wilmington, known as the Penjerdel (from *Pennsylvania, New Jersey,* and *Del*aware) Region,[38] some preliminary studies of possible government organization to manage air quality for the region have been made by the Fels Institute of the University of Pennsylvania. The adaptations of the interstate compact devised by the Fels Institute study team are worth some attention, as they represent one of the first attempts to face the problem on an interstate basis.

The study team suggests an interstate agency, created by interstate compact but containing a regional constituency representation, to "plan, manage, and control" the Penjerdel airshed. The power and duties of the compact agency would include:

1. undertaking research;
2. establishing, promulgating, and enforcing regulations, including air quality standards and emission standards;
3. designating control districts;
4. issuing and enforcing orders against offending emissions;
5. establishing warning systems and exercising emergency powers;
6. requiring the registration and reporting of emissions;
7. requiring and issuing licenses and permits to emission sources.[39]

These are seen to be a mixture of executive and policy powers. In particular, the quality and emission standards by districts constitute the heart of the policy on social choice issues involved.

Recognizing the deficiencies of the one-state/one-vote principle as applied to this kind of problem, the study group suggested that the compact agency (commission) be composed of two members from each state (Pennsylvania, New Jersey, and Delaware), one representing the government of the state and the other that part of the state located in the Penjerdel region.

---

[38] Area of about 4,454 square miles with about 5 million people.

[39] Taken from "Governmental Organization for a Regional Air Resource Management and Control System" (Fels Institute, University of Pennsylvania, 1968). It should be noted that the study, done under contract from HEW, was responsive to the HEW definition of the problem, as it should have been. My critique is somewhat unfair since I am suggesting a course of action beyond the scope of their study.

They further suggested that the local representative in each state be elected by the people from the counties concerned. Thus, local political parties could become involved.

One provision sets up a dual voting structure somewhat similar to that of the Security Council of the United Nations: that "no action . . . shall be taken . . . unless a majority of the membership, including . . . a majority of the State representatives, shall vote in favor thereof." This provision practically weights the states' votes double those of the local representatives, although nominally it is a three-to-two weighting. (That weighting which ensures that no majority exists unless at least two states concur, *and* ensures that at least one locality concurs—three states cannot pass bills alone—is a three-to-two weighting with a two-thirds majority decision rule.) The nominal voting power may hence be calculated as if each state had three votes and each local representative had two, with ten votes needed to pass:

Votes	Number of voters	Total number of votes	Shapley value (%)
3	3	9	21.67 (each)
2	3	6	11.67 (each)
	6	15	

two-thirds majority = 10

Table 10 shows the power of different bloc arrangements. Thus, the best that the people involved can have is one-half the voting power, and that only under the highly improbable assumption that the region is totally united and the states totally fragmented.

The voting arrangements are, however, a minor matter in this case, since there are more serious flaws in the proposal. The right to control air quality and emission standards over the area, were it to be given to the proposed agency, would have great leverage over the industrial development decisions for the area and for land use generally. The framework for making these decisions (or, more precisely, for placing restrictions on the decisions of others) should obviously not be that of an air quality control commission; allowing it to decide these issues is analogous to letting the Port of New York Authority continue to make transport policy.

Some attention to the necessity for coordination with the water authority

Table 10. The Power of Penjerdel Voting Blocs Measured in Shapley Values

State and local representations	Votes	Number of voters	Shapley value (%)
$S_1 + L_1 + L_2$	7	1	67
$S_2 + L_3$	5	1	17
S_3	3	1	17
$S_1 + L_1$	5	1	33
$S_2 + L_2$	5	1	33
$S_3 + L_3$	5	1	33
$S_1 + S_2$	6	1	60
S_3	3	1	10
$L_1 + L_2 + L_3$	2	3	10 (each)
$S_1 + S_2 + S_3$	9	1	75
$L_1 + L_2 + L_3$	2	3	8 (each)
$L_1 + L_2 + L_3$	6	1	50
S_1	3	1	17
S_2	3	1	17
S_3	3	1	17

Note: Majority = 10; S = state representatives, L = local representatives.

in a region was given by the study team. The issue, however, is not just coordination of residuals management policy, but rather the question of who is authorized to make decisions about the size, shape, land uses, and relative economic as well as physical health of the Penjerdel region. The precision with which an interstate compact on air management is drawn pales beside the social choice issues involved. These are better discussed in the framework of a metropolitan area government of general jurisdiction, rather than in a special-purpose agency with a very limited pattern of representation and little scope for party politics.

Minneapolis–St. Paul Metropolitan Council

In 1967, the state legislature of Minnesota created "an administrative agency" for the purpose of coordinating the planning and development of the metropolitan area comprising the counties of Anoka, Carver, Dakota, Hennepin, Ramsey, Scott, and Washington.[40] The agency, the Metropolitan Council, has as its members fifteen appointees by the governor, fourteen

[40] Area of about 3,000 square miles with approximately 2 million people.

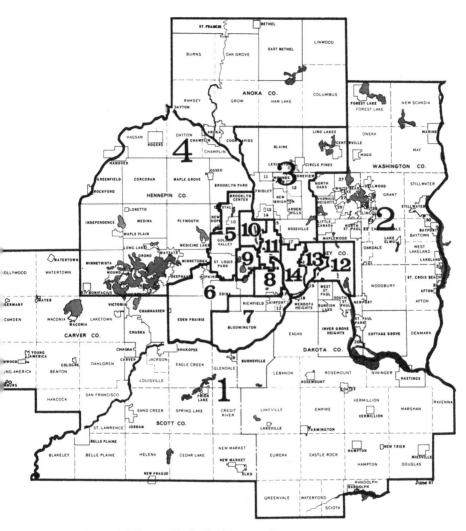

Figure 2. Minneapolis–St. Paul Metropolitan Council districts.

of whom come from equal-population districts (see figure 2) making up the area, and the fifteenth designated as chairman. The motion to have the fourteen elected lost by the narrowest of margins, for, though deemed an administrative agency, the council is, and increasingly will be considered, the policy-making body for the area, making the environmental choices for the region.

The council's present powers include:

1. preparation and adoption of a development guide for the metropolitan area;
2. review of all long-term plans of each independent commission, board, or agency and government unit in the metropolitan area if the plans have area-wide or multicommunity effect, or have a substantial impact on metropolitan development;
3. review of applications of all governmental units, independent commissions, boards, or agencies operating in the metropolitan area for loans or grants from the United States or any of its agencies, which require review by a regional agency;
4. right to levy a tax not to exceed 7/10 mill of assessed valuation of taxable property in the area to provide funds for its operation (estimated to yield over $1 million in 1970);[41]
5. development of a data center;
6. research on areawide problems including air pollution, parks and open space, water pollution, solid waste, and tax structure of the area and consolidation of common services.[42]

Some of the independent agencies over which the council exercises policy control (for review in most instances means power to reject) are the Minneapolis–St. Paul Sanitary District, the Minneapolis–St. Paul Metropolitan Airports Commission, the Metropolitan Mosquito Control District, the Metropolitan Transit Commission, the Hennepin County Park Reserve District, the North Suburban Sanitary Sewer District, hospital districts (North Suburban and Forest Lake), the Dakota-Scott Library District, the Lake Minnetonka Conservation District, soil conservation districts in six of the seven counties, four watershed districts, Metropolitan Park Board, and the Metropolitan Zoo Board. These are, of course, in addition to the regular municipal governments in the area. The reality of the review process is evoked by the fact that the council denied the Airports Commission's choice of a new airport site on the grounds that it might

---

[41] The council may also issue general obligation bonds for the acquisition and betterment of sewer and treatment works.

[42] Taken from "Referral Manual" (Metropolitan Council of the Twin Cities Area, March 1968), and other official statements of the Metropolitan Council.

interfere with the recharge of reservoirs that supply part of the area's water supply.[43]

It was after considerable discussion inside the legislature and through the very active and prominent Citizen's League that the unusual step of subordinating (but not abolishing) the independent boards was taken. It was clearly recognized that the independent boards are mostly operating agencies. Rather than replace them with professional staffs, it was decided to allow them to be intermediate lay boards between the professional staffs of their own special-purpose agencies and the multipurpose Metropolitan Council. In most, but not all cases, the independent board members are appointed by the chairman of the Metropolitan Council.

The practical and useful result is that the council is relieved of the multitude of operating decisions in many special fields, but retains policy control over the special-purpose agencies. At the same time, the kind of executive agency accommodation of interests *at the implementation level* can be effectively handled by the appointed special-purpose boards. In practice, there are some problems in sorting out the roles of general-purpose council *staff* versus special-purpose board *members*. These will not be solved, as the participants tend to believe, by "working it out"; there is an inevitable conflict here that the council members, particularly the chairman, will have to contain and channel to productive purpose.

The council represents the first real innovation on the metropolitan government problem since World War II. Its chief virtue is that it has broken out of the council of governments (COG) mold, which suffered not only from its one-town/one-vote (implicit) principle, but also from its total lack of authority to act in most instances. The council is not attempting to supplant municipal governments in the area, nor does it draw its strength from, or depend upon, them. It enjoys widespread support of municipalities. It is at the state level that some unease is felt about the council and that unease, plus some second thoughts of state legislators, represents the only potential danger to the eventual consolidation of the council as an elected, multipurpose government.

It can be seen that a Bay Commission or an Air Pollution Commission

[43] I am indebted to Ted Kolderie, executive director of the Minneapolis–St. Paul Citizen's League, and his staff for information on this point and for their help on this section. Neither Mr. Kolderie nor the league is, of course, responsible for my analysis or conclusions.

could fit easily under such an areawide governmental unit. It is also clear that, were a multistate area involved, the concerned states could combine powers to create a multistate metropolitan council. The interstate compact device is not limited to setting up one kind of organizational structure.

The Metropolitan Council of the Twin Cities area, unlike the Bay Commission or the Air Pollution Commission, does offer a general pattern worthy of emulation. It fits into the system of government both in terms of law and of utility. It will offer an excellent opportunity for the two political parties to contest for seats when (if) the commissioners become elected. It has taxing and limited bonding power and the potential of getting more. It provides, most importantly, that metropolitan stage essential to creative action. Norton Long has put it most eloquently:

> To provide a leadership that can appreciate the desirability of a rich community life, a significant theatre of action and the means to significant action are requisite. Such a theatre of action exists in potentiality in our metropolitan areas if they can be given the political form requisite to the recruitment and functioning of a metropolitan leadership. Such a leadership is essential if local self-government is not to atrophy in the decline of the central city, the triviality of the suburb, and the growth of *ad hoc* and upstream agencies for the administrative, piecemeal handling of urgent problems. There is still no substitute for territorial representation as a means to coordinate and integrate the functional organizations that share a territorial field. Unless the means of electing and instrumenting such a territorial leadership can be found, local self-government will give way to the administration of people rather than the self-direction of citizens.[44]

There are many ways in which a multipurpose government can operate to change the rules under which people and municipalities compete. The movement toward sharing of some part of the property tax is well underway in the Twin Cities area. The issue here is not to pry some of the local revenue away from municipalities, but to share new tax revenues resulting from new business and industrial investments, relieving every municipality from the necessity of competing for new industry.

As industry is attracted to the economic area, not usually to a particular site within it, an areawide approach can increase benefits for all without reducing local options for differentiation on municipal services. While it is

[44] Norton E. Long, "Citizenship on Consumership in Metropolitan Areas," *Journal of the American Institute of Planners* 31, no. 1 (1965): 6.

true that the more densely populated areas will dominate the council, and thus sparsely settled, outlying areas will come more under the control of the center, it is wholly appropriate that this should happen. The existing practice, wherein windfall gains are reaped by landowners outside the taxing power of the city whose economic base is responsible for the gain is, while common, quite intolerable on any grounds of equity or efficiency.

The creation of the council may enable some intermediate levels of government—the county, for example—to reduce their functions and in some cases to disappear. In other contexts than that of the Twin Cities, the central city may gradually disappear as a unit, no longer appropriate for decision making, leaving smaller units to handle purely local affairs and an area-wide unit to handle metropolitan problems.[45]

The lesson here is simply that the appropriate size for general-purpose governments may change and has changed over time. When this happens, the governments become ineffective, not because they are general-purpose governments composed of elected officials, but because they no longer provide a stage on which leadership can play. Because that stage is at two levels in urban areas today—the neighborhood level and the metropolitan level—this is where general-purpose governments can be useful.

At the risk of belaboring the point, it is useful to recall here that with a number of independent issues to consider, the council members can make trade-offs over the whole spectrum of issues. They do not have to compromise each special issue to the median position in order to resolve it, neither do they have to let every area have *some* transit, *some* parks, or whatever. The systems can each be designed more efficiently overall and the trades serve intensity of interest concerns. In other words, the utility mechanism can work.

CONCLUSION

Viewing environmental quality as a social choice problem and representative government as the utility mechanism that has unique capabilities for aggregating individual preferences into social choices enables me to make

---

[45] This is consistent with the recommendations of two recent studies of metropolitan government. See Committee for Economic Development, *Reshaping Government in Metropolitan Areas* (New York: CED, February 1970); and Alan K. Campbell, ed., *The States and the Urban Crisis* (Englewood Cliffs, N.J.: Prentice Hall, 1970).

a number of general conclusions about the present state of institutions in the environmental field. These may be summarized as follows:

1. There is a need to redress the balance between the legislative and the executive role in environmental choices at all levels of government. Historically, we may be in a period akin to the early eighteenth century in North America, at which time the states and commonwealths seized policy control from crown-appointed governors and executive councils. The supremacy of the legislature for policy determination (the old Whig position), which was firmly established at that time, has gradually eroded over the centuries. We now face, again, the need for strong legislatures. Reestablishing them will require sweeping out many twentieth-century habits of legislatures, among them the excessive dependence on seniority and the abuses of the committee system, and creation of some twentieth-century research capabilities within the legislatures.

2. There is a need to force environmental issues into partisan politics at every level of government. There is, unfortunately, a trend in the opposite direction, with both parties adopting pious statements but leaving the solution of any controversial issues to administrative agencies that are not subject to citizen pressures.

3. There is a need to focus directly on institutional design. In doing so, we must recognize that making social choices between hard alternatives requires technical expertise and political expertise. Both are necessary; neither is sufficient by itself. So long as we have our present system of government, we must recognize that it was designed to make social choices through legislative action. We must, therefore, disenthrall ourselves from the habits of applying interstate compacts on the rigid one-state/one-vote principle and of copying business organization and forms.

Using the tools of utility analysis, we can examine the political processes and outcomes of present and proposed political institutions, judging them in terms of:

    (a) the territorial reach of their representation and of the problems they face;

    (b) the number of independent issues under their control;

    (c) the decision rules

        (i) relative to passing and blocking measures that come before them,

      (ii) relative to what utility "game" the situation forces the interested parties to play;

  (d) the viability of party politics in the institution.

4. Environmental issues, while they have local, state, and federal aspects, are often primarily regional in nature. Sometimes the region is in one state, sometimes in more than one. There is therefore a need for representative governmental structures at the regional level. These may be created either by one state or by several acting through the compact route. In either case, both the legislative and executive dimensions must be attended to. Both dimensions need not be in the same agency. In the Minneapolis–St. Paul case, for example, they are not.

5. While not a major subject in this paper, the utility framework used leads inescapably to the conclusion that the public works committees in Congress have become almost wholly inappropriate mechanisms for making project decisions on water and other environmental issues. The point here is not that pork-barrel politics is bad or that the men on these committees are wicked. It is simply that the physical environment is now too important to be used as the principal, or even a major, currency of Congress. The need is not, therefore, to reform the public works committees, but to remove water projects and other regional project investment concerns from Congress. This process has, in fact, already started. It will be hastened as congressmen learn of their constituents' resistance to certain projects in their districts. Dams, army bases, and all the other "prizes" awarded are no longer greeted with unmitigated joy. This is a healthy trend in utility terms, as it will hasten the move to let local citizens make local decisions. In passing, it should be noted that the widest possible scope for trading would still remain with the national Congress. Many of these trades are regional in nature and can take the place of water as a medium of exchange. Moreover, the executive side of the federal government, in particular the Corps of Engineers, could easily put its expertise to the service of, and under the policy direction of, regional or state agencies on a project-by-project basis.

The federal role would then emphasize, on the legislative side, the questions of how much money should be appropriated for environmental purposes and what regional distribution should be made. On the executive side, the federal role would emphasize the availability of technical expertise in many areas on which regions and states could call. The states, regions, and localities would, on the other hand, be making environmental choices (in a few cases being limited by nationwide minimum standards), and deciding how to allocate federally disbursed monies, what additional monies were needed, and how best to implement these decisions. The picture this presents is far different from the one that now exists, but it is far closer to the process of self-government envisioned for the country two centuries ago.

3

Social Choices and Individual Preferences: Is There a Connecting Mechanism?

INTRODUCTION

When decisions are forced into collective choice mechanisms, it is not surprising that the decisions are colored by other decisions also in process of being made there. We are, collectively, concerned about welfare, and hence we may use decisions about environmental quality to advance a particular income redistribution scheme, say, improving water quality in a stream because poor blacks use it for swimming and fishing. We are, collectively, concerned about regional economic "balance" and may use decisions about environmental quality to advance a particular region, say, tying a flood control project in with a water navigation scheme to bring "low-cost" water transport to an area previously served only by rail, air, and road.

When we mix our motives for public investment, we are often subject to the apt criticism that our instruments are inefficient for the purposes. Perhaps swimming pools and food stamps can produce equal benefits to poor blacks at less cost than improved water quality. No doubt rail rates can be forced down more cheaply than by digging what are, in effect, canals for

Reprinted, with minor revision, by permission from J. R. Conner and E. Loehman, eds., *Economics and Decision-Making for Environmental Quality* (Gainesville: University of Florida Press). Copyright 1973 by the State of Florida Board of Trustees of the Internal Improvements Fund. The paper was originally presented at the Seminar on Economics and Decision-Making for Environmental Quality, University of Florida, 23 February 1971.

barges. Much of the criticism is well founded, on efficiency grounds, but we must not allow it to obscure the necessity for the collective choice on environmental goods, even if we ignore income redistribution and equity considerations. Environmental quality partakes of the characteristics of a public good, particularly in its nondivisibility aspects, so that we cannot escape making collective choices about it.

How are we to decide how much environmental quality we are to have? One obvious answer is, as much as we are willing to pay for. But how are we to determine that if we cannot divide it up and price it in individual, separate packages? Moreover, since many different bundles of environmental quality might cost the same amount but have far different patterns in the distribution of the costs and benefits, how are we to distinguish one bundle from another? And if you view motorboating as recreation and I view it as noise pollution, how are we going to decide simple inclusion and exclusion in our definition of environmental quality?

Perhaps we can agree that both economic and political beliefs have their roots in personal utility and use that as the bedrock of our analysis. Then we will be able to think about my disutility and your utility associated with motorboats. How do we combine the two? We have known since Bergson that we cannot sum utilities across individuals. We are beginning to discover that we cannot sum benefits across individuals (except by ignoring welfare implications). How are we to relate personal utilities (preferences) to the social choices we make about the environment?

FOUR CASES

In this paper I intend to examine the ways by which we attempt to make the connection between personal preferences and social choices. To make the discussion concrete I will examine four specific situations involving water quality: two actual cases, the Delaware and the Ruhr; and two hypothetical cases, the Dorfman-Jacoby model of river basin management and my own work. (Needless to say, this introduces a certain bias in the discussion which the reader must guard against.) I shall be concerned specifically with the setting of standards or levels of water quality and thus will be neglecting a whole host of technical issues, in particular, the implementation features in each case. Since the focus is to be on *how* the decision to adopt a certain goal is made, I will not emphasize how the goal is to be achieved except when this is intertwined with the setting of the goal itself (as it is in cost-sharing arguments). Each case examined raises specific questions about the connection between individual preferences and social choices. Discussion of these questions will be undertaken in a later section.

The Choice of Water Quality in the Delaware Estuary

A detailed and critical review of the steps leading to the formation of the Delaware River Basin Commission (DRBC) is not warranted here, nor am I informed enough to make it. An analysis of the studies made on the Delaware River concerning water quality is made by Kneese and Bower,[1] and their analysis will be relied upon for our purposes.

The social choice about water quality in the Delaware estuary was related directly to preferences by three mechanisms. In point of time the first of these was the system of advisory committees set up as a part of the Delaware Estuary Comprehensive Study (DECS) undertaken by federal agencies at the request of the concerned states and interstate agencies before the commission was established. In the words of the present director of the DRBC:

> The Delaware Estuary Comprehensive Study selected various technical committees which would permit the voice of the estuary community in the development of recommended judgments. The committee on water use included four groups—one represented recreation, encompassing conservation interests in fish and wildlife; another was representative of the general public; a third represented industry; while the fourth was made up of representatives of local governments and planning agencies.

The study was a notable technical advance over earlier studies of water quality and included the development of a model for looking systematically at the interrelationships among reaches of the river and for arriving at cost-minimizing solutions under different sets of assumptions.

One of the outputs of the DECS was the identification of "five 'objective sets,' each representing a different package and spatial distribution of water quality characteristics, with the level of quality increasing from set 5 (representing 1964 water quality) to set 1."[2] Total and incremental costs associ-

[1] Allen V. Kneese and Blair T. Bower, *Managing Water Quality: Economics, Technology, Institutions* (Baltimore: The Johns Hopkins Press for Resources for the Future, 1968).

[2] Kneese and Bower, *Managing Water Quality*, p. 226.

Table 11. Costs and Benefits of Water Quality Improvement
in the Delaware Estuary Area

(million dollars)

Objective set	Estimated total cost	Estimated recreation benefits	Estimated incremental cost		Estimated incremental benefits	
			Mini-mum[a]	Maxi-mum[b]	Mini-mum[a]	Maxi-mum[b]
1	460	160–350				
			245	145	20	30
2	215–315	140–320				
			130	160	10	10
3	85–155	130–310				
			20	25	10	30
4	65–130	120–280				

Source: Allen V. Kneese and Blair T. Bower, *Managing Water Quality: Economics, Technology, Institutions* (Baltimore: The Johns Hopkins Press for Resources for the Future, 1968), p. 233.

Note: All costs and benefits are present values calculated with 3 percent discount rate and twenty-year time horizon.

[a] Difference between adjacent minima.

[b] Difference between adjacent maxima.

ated with each set were estimated and benefits associated with moving from one set to another were estimated. Table 11 gives a summary of these costs and benefits.

The water-use advisory committee recommended adoption of objective set 3, indicating an agreement with the incremental costs and benefits estimated by DECS. The incremental benefits relate chiefly to recreation benefits, specifically swimming, boating, and sport fishing, since there were no benefits of any magnitude to industry and municipalities. Indeed, industry might be expected to suffer negative benefits (higher costs) from corrosion stemming from higher oxygen levels in objective sets 3, 2, and 1.

The second mechanism that related preferences to the social choice under consideration was the series of public hearings held by the DRBC prior to commission adoption of quality standards. Hearings were held in Delaware, Pennsylvania, and New Jersey, those in the latter two being the most extensive in terms of testimony. Kneese and Bower note that "nearly all discussion related to objective sets 2 and 3; the other sets were recognized by almost everyone as not being in the public interest."[3] While I agree with that statement, I do also note for the record the statement of the alternative commissioner for New Jersey at the beginning of the hearing: ". . . New Jersey advocated and the Commission adopted that the public hearing on these water quality standards . . . be presented in the form of two alternatives representing the two depictive sets deemed feasible for consideration at this time—objective sets 2 and 3."[4]

The pattern of testimony at the hearings found most industries willing to support objective set 3, though they carefully noted that no advantages accrued to them by this choice (they would have been happy with sets 4 and 5, which cost less); the representatives of governments supported either set 3 or set 2, depending mainly on whether they were elected or appointed. The appointed (executive) official was more apt, of course, to be concerned with the costs of responding to the higher quality level, whereas the elected official clearly saw votes in the advocacy of "clean" water. Conservation groups supported set 2, but conscientiously pointed out that it was sinful not to go for objective set 1, which they really preferred. The water-use committee established under DECS testified, as did its individual subcommittees, and continued to support set 3. One subcommittee (fish and wildlife) jumped the reservation in the Delaware state hearings, advocating set 2.

The third mechanism for relating preferences to the social choice was the act of choosing itself. The commission, which chose standards falling between set 3 and set 2, is an agency set up by interstate compact. The commissioners vote as sovereigns; that is, each state has one vote. There is thus some connection, however tenuous, between the vote and what the governor of each state assesses (or chooses not to assess) as the preferences of the people in his state.

The questions raised by the process of decision making in the Delaware case are threefold:

1. Is the benefit-estimating technique valid in this context?
2. What purpose is served by using representatives of interest groups, including one "representing" the general public, to advise in the formation of public policy?

[3] Kneese and Bower, *Managing Water Quality*, p. 233.
[4] Delaware River Basin Commission Water Quality Hearings, Trenton, New Jersey, 26 January 1967, pp. 13–14.

3. How does the one-state/one-vote tradition of interstate compacts affect outcomes on social choice issues?

Water Quality in the Ruhr

The German *Genossenschaften* are justly famous for their long and distinguished record for management of basinwide water systems, a record that includes both technical and economic innovations of a high order.[5] In Fair's words, "One of the river-basin authorities, the Emschergenossenschaft, has been in action, almost without constitutional change, for over half a century, under no less than four sovereignties: a constitutional but authoritarian monarchy, a strongly centralized federalist republic, a ruthless dictatorship, and a decentralized federal republic. Survival has, patently, been through strength."[6]

The *Ruhrverband*, the *Genossenschaft* responsible for water quality in the Ruhr, is described by Kneese and Bower:

> The basic political power in the Ruhrverband lies in the governing board which is made up of: (1) the owners of business and industrial establishments and other facilities that lie in the Ruhrverband area and that contribute to water quality deterioration in the Ruhr or its tributaries, and those who benefit from the activities of the Ruhrverband to the extent that they make a specified minimum financial contribution to its activities; (2) communities within the Ruhrverband area; and (3) the Ruhrtalsperrenverein as a representative of the waterworks and other water withdrawal facilities.
>
> The political organs are the assemblies and the board of directors. The assembly members (about 1,500) elect the board of directors, approve or disapprove the plan for water quality management, approve or disapprove the assessment of charges, and decide upon the basic method for calculating the level of charges. The assembly reaches its conclusions on the basis of absolute majority with the number of votes cast by each member being dependent upon the amount of financial contribution.[7]

Since the assessments levied by the *Ruhrverband* have the legal force of taxes and the level of assessments relates directly to the water quality goals agreed upon, we obviously have a social choice issue at hand, but a mech-

---

[5] I have relied on Gordon Fair, "Pollution Abatement in the Ruhr District," in Henry Jarrett, ed., *Comparisons in Resource Management* (Baltimore: The Johns Hopkins Press, 1961), and on Kneese and Bower, *Managing Water Quality*, for my understanding of the Ruhr water-management system.

[6] Fair, "Pollution Abatement," p. 167.

[7] Kneese and Bower, *Managing Water Quality*, p. 261.

anism somewhat different from those in the Delaware case. In some respects, the mechanism resembles a mixture of a private corporation and a cooperative. The voting basis has much of the Wicksellian theme of tying specific expenditures directly to specific taxation. Fair notes that "both the cost of pollution abatement and the value of direct, as well as indirect, benefits derived by a member . . . are assessable" in some of the *Genossenschaften.*[8] Thus the incidence of benefits as well as costs can be calculated and used as an element in the assessment of members.

It is hard to assess this German mechanism without getting entangled in differences between cultures. For instance, would the charges levied by such a body against a municipality (which would be a member) and hence passed on to taxpayers stand against a "due process" argument in this country?

If one ignores the questions of direct transferability of the mechanism, however, and views it only as a mechanism for relating social choice to preferences, the following questions arise:

1. Is it possible that some pay for "benefits" they do not want? Or is there a bias for underproduction of benefits?
2. Are there "interests" in addition to those represented?
3. How would a shift in public tastes toward water quality be reflected by the *Ruhrverband*?

The Dorfman-Jacoby Model of River Basin Management

Robert Dorfman has recently outlined a conceptual model "within which the political, economic, and technological aspects of regional water management can be brought together." With Henry Jacoby, Dorfman then constructed a hypothetical river valley and an authority (the Bow Valley Water Pollution Control Commission) in order to test the model.[9] In the following account only the political aspects of the Dorfman model will be discussed. (The discussion should *not* be relied upon for an understanding of the Dorfman-Jacoby work.)

Dorfman views a river basin authority as a "synthesizer of the goals and objectives of other groups and agencies" and thus faces directly the question of translating preferences into social choices. His model accepts will-

[8] Fair, "Pollution Abatement," p. 152.

[9] Robert Dorfman, Henry Jacoby, and Harold A. Thomas, Jr., eds., "Models for Managing Regional Water Quality," mimeographed (Harvard University, 1970), see chaps. 2 and 3.

ingness to pay as a measure of benefit, while recognizing the difficulties of application. Assuming reasonable estimates exist, Dorfman then defines a class of feasible decisions as Pareto-admissible decisions as follows:

> A decision Y is said to be Pareto admissible if it is feasible (i.e., if it satisfies all pertinent constraints) and if there does not exist any alternative feasible decision X for which $NB^i(X) \geq NB^i(Y)$ for all participants i, with strict inequality holding for at least one participant. [NB = net benefits][10]

The purpose of the model is, first, to determine the set of Pareto-admissible decisions under any given set of constraints. Some of these constraints may be the powers (and lack thereof) of a river basin authority. Thus some idea of how changes in the constitution of the authority affect the set of admissible decisions may be gained.

Functioning of the model involves, of course, an aggregation of benefits and costs. This creates problems, not of estimation alone but also of interpersonal comparisons of utility. (Dorfman's discussion is clear on these issues and is highly recommended.) His solution is the traditional one of assigning weights, w_i for each participant (municipality, industry, and so forth) whose costs and benefits are at issue. These weights he calls political weights and they are used to form the objective function

$$W = \sum_i w_i \, NB^i(X)$$

that is to be maximized with respect to X, assuming X is a feasible solution. Each choice of a set of w_i's will produce a new Pareto-admissible decision. Since there is an infinite number of w_i sets, Dorfman's procedure at this point is to solve for "a number of widely different sets of political weights."

For any specified economic and technical constraints, and with a specified set of political weights, the feasible and Pareto-admissible decision is generated. By changing weights, a number of admissible decisions are generated. A priori there is nothing known about whether different assumed weights will give widely different decisions or the same decision. Dorfman is well aware that the political weights must be assigned on the judgment of someone or other (i.e., that they are not known), but he suggests that reasonable guesses may be made. Moreover, he suggests that if the range of

[10] Ibid., p. 30.

decisions proves to be fairly narrow—that is, not sensitive to the choice of weights—then the model's predictive powers may indeed be strong.[11]

The latter point is the main point of the application of the model to a hypothetical case by Dorfman and Jacoby. Their illustration affords an excellent explication of the whole model, but I will again confine my discussion to the political (translating preferences to choices) aspects of the illustrative case.

The hypothesized Bow River Valley (see fig. 3) contains an industry, two cities, and one recreation facility. The water quality parameter subject to control is dissolved oxygen (DO), and a minimum level (3.5 ppm) is required at the state line. The control commission may choose this level or any feasible level above this minimum and may implement its decision by regulating the level of treatment required at each source of pollution—the cannery and the two municipalities. The water quality problem hypothesized is an upstream-downstream conflict. Beyond Bowville the stream becomes anaerobic, due to the discharges from the town and from the cannery; hence improved treatment by them benefits Plympton. Various constraints are imposed having to do with politically feasible tax increases in the two towns, the amount of improved treatment the cannery can bear without going out of business, public interest in improving the river, and so forth. A schedule of quality standards similar to the objective sets in the Delaware case is supplied—classes A through D in terms of decreasing DO levels. The Bow Valley Water Pollution Control Commission is composed of members from the city councils of both cities and the deputy state commissioner of parks and recreation; it is chaired by the crusading editor of the *Bow Valley News*. Precise numbers or decision rules are not specified.

The model was solved for seventeen different weight allocations (see table 12) and three quality minima (B through D). In order to give a measure of national interest, weights were assigned to the Federal Water Quality Administration (FWQA). For example, the first weight allocation (0, 0, 0, 1) gives exclusive emphasis to national, as opposed to local, interests, in the traditional cost-benefit sense. In addition, least cost (aggregate) solutions

---

[11] The usefulness of the model in excluding non–Pareto-admissible decisions is also claimed, but this "usefulness" depends on whether or not the weights have been "correctly" judged.

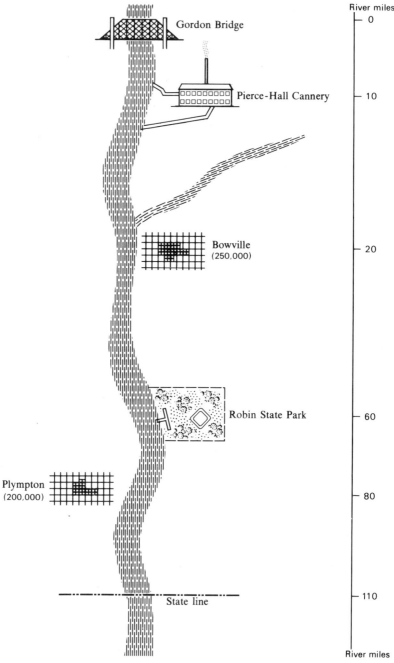

Figure 3. Main features of the Bow River Valley (from Dorfman, Jacoby, and Thomas, eds., "Models for Managing Regional Water Quality").

Table 12. Weight Allocations Used to Explore Possible Decisions

Decision number	Relative weight assigned to			
	Pierce-Hall w_1	Bowville w_2	Plympton w_3	FWQA w_4
1	0	0	0	1
2	3	1	3	3
3	3	3	1	3
4	3	3	3	1
5	1	3	3	3
6	4	1	4	1
7	1	4	1	4
8	4	4	1	1
9	1	1	4	4
10	1	4	4	1
11	1	5	3	1
12	1	3	5	1
13	1	6	2	1
14	1	2	6	1
15	1	1	7	1
16	1	7	1	1
17	7	1	1	1

Source: Dorfman, Jacoby, and Thomas, eds., "Models for Managing Regional Water Quality."

were calculated for each of the *B* through *D* standards. These solutions appear as decisions 3, 7, and 11.

The total number of Pareto-admissible solutions generated by the 17×3 trials was 14. These are displayed in table 13.

Tables 14, 15, and 16 display costs and benefits associated with the achievement of water quality minima. Net costs refer to the outlays required to meet the treatment levels specified in table 13. The net benefits refer to the difference in benefit between any decision and decision 3, the least-cost plan to achieve class *D* water. Note that the seventeen weighting schemes converge on a substantially smaller number of admissible decisions in each table.[12]

Some of the questions that are raised by the Dorfman model include:

1. What is political about his political weights?

[12] Nine solutions using class *D* as a minimum, eight solutions using class *C*, five solutions using class *B*. Some solutions appear in more than one class minimum.

Table 13. Decisions Recommended by the Analysis

Decision number	Quality class achieved	Treatment level in terms of percentage BOD removal by		
		Cannery	Bowville	Plympton
1	D	95	50	67
2	D	90	52	67
3	D	80	55	67
4	C	95	65	61
5	D	90	66	61
6	C	90	67	61
7	C	80	70	60
8	B	95	80	54
9	B	94	80	54
10	C	90	80	55
11	B	90	81	54
12	C	80	80	56
13	B	80	85	54
14	B	90	90	51

Source: Dorfman, Jacoby, and Thomas, eds., "Models for Managing Regional Water Quality."

2. What are the limitations on his definition of Pareto-admissibility?

3. What is the predictive value of this model?

A Utility Model for Aggregating Preferences

My own work has been concerned with the direct confrontation of the aggregation problem of individual preferences into social choices. Thus I have had to contend with the General Possibility Theorem as restated by Kenneth Arrow in his landmark work, *Social Choice and Individual Values.* Elsewhere (see appendix C) I have attempted to show that a two-party system could, under majority rule, produce decisions on issues that would be identical to the decisions produced if all individuals could take advantage of vote trading. This result uses Arrow's neglected Possibility Theorem for Two Alternatives and provides a set of rules for choosing the two alternatives. The rules turn out to be a simulation of a perfectly functioning two-party system.

The basic building block of the utility model is a vector of preferences related to a given set of independent issues, for example, $\begin{bmatrix} Y_2 \\ N_1 \\ Y_3 \end{bmatrix}$, which combines yes-no voting stances with an ordinal ranking of the importance of

Table 14. Decisions Corresponding to Specified Weight Allocations under Minimum Quality Classification D

Weight allocation	Decision number	Quality class achieved	Net cost ($ thousand/yr)				Net benefits ($ thousand/yr)			
			Pierce-Hall	Bow-ville	Plymp-ton	FWPCA	Pierce-Hall	Bow-ville	Plymp-ton	FWPCA
1711, 1621	1	D	95	201	302	829	−87	72	−3	−61
4411	2	D	35	217	302	750	−27	48	−2	10
L. C.	3	D	8	250	301	746	0	0	0	0
3313, 1531, 1414, 1441, 0001	5	D	35	349	255	863	−27	−38	160	101
3133, 1144, 1351, 1333, 3331	10	C	35	491	204	985	−27	−131	282	147
1261	11	B	35	512	199	1007	−27	−148	292	139
7111	12	C	8	491	214	953	0	−158	255	125
4141	13	B	8	568	198	1034	0	−219	294	97
1171	14	B	35	659	169	1162	−27	−279	348	36

Source: Dorfman, Jacoby, and Thomas, eds., "Models for Managing Regional Water Quality."

Table 15. Decisions Corresponding to Specified Weight Allocations under Minimum Quality Classification C

Weight allocation	Decision number	Quality class achieved	Net cost ($ thousand/yr)				Net benefits ($ thousand/yr)			
			Pierce-Hall	Bow-ville	Plympton	FWPCA	Pierce-Hall	Bow-ville	Plympton	FWPCA
1711 1621 4411	4	C	95	344	251	952	−87	−21	169	34
3313 1414 1531	6	C	35	360	251	873	−27	−45	170	105
L. C. 1441 0001	7	C	8	393	249	869	0	−94	171	94
3133 1144 1351 1333 3331	10	C	35	491	204	985	−27	−131	282	147
1261	11	B	35	512	199	1007	−27	−148	292	139
7111	12	C	8	491	214	953	0	−158	255	125
4141	13	B	8	568	198	1034	0	−219	294	97
1171	14	B	35	659	169	1162	−27	−279	348	36

Source: Dorfman, Jacoby, and Thomas, eds., "Models for Managing Regional Water Quality."

Table 16. Decisions Corresponding to Specified Weight Allocations under Minimum Quality Classification B

Weight allocation	Decision number	Quality class achieved	Net cost ($ thousand/yr)				Net benefits ($ thousand/yr)			
			Pierce-Hall	Bow-ville	Plymp-ton	FWPCA	Pierce-Hall	Bow-ville	Plymp-ton	FWPCA
1711 1621	8	B	95	487	200	1074	−87	−115	292	80
1441 1531	9	B	81	491	200	1056	−73	−121	292	96
4411 3313 0001 3133 also 1144 L.C. 1351 solu- 1333 tion 3331 1261	11	B	35	512	199	1007	−27	−148	292	139
7111 4141	13	B	8	568	198	1034	0	−219	294	97
1171	14	B	35	659	169	1162	−27	−279	348	36

Source: Dorfman, Jacoby, and Thomas, eds., "Models for Managing Regional Water Quality."
Note: 1414 not listed.

the issues to any individual. Thus, above, the individual is for the first issue, against the second, and for the third. The second is most important to him, the first next, and the third issue least important. A display of three such vectors (a three-man committee) might give us (P = pass, F = fail):

$$Y_2 \quad N_1 \quad Y_1 - P$$
$$N_1 \quad Y_3 \quad Y_3 - P$$
$$Y_3 \quad Y_2 \quad Y_2 - P$$

in which case, if we sum the vote across rows, all three issues would be passed by the committee (assuming majority rule). Note, however, that the first two men can trade votes on the top two issues:

$$
\begin{array}{ccc}
 & \textcircled{N_2} & N_1 & Y_1 - F \\
Y_2 \quad N_1 & & \\
\diagdown \diagup & \text{giving} \quad N_1 & \textcircled{N_3} & Y_3 - F \\
N_1 \quad Y_3 & & \\
 & Y_3 & Y_2 & Y_2 - P
\end{array}
$$

making both better off. The outcome on the three issues now is that the first two fail and the third passes. The two men have improved their positions (both are better off) at the expense of the third, who is now worse off. Clearly we are not dealing with Pareto states here but with an exchange mechanism that has third-party effects. (In larger matrices, some non-traders gain and others lose.)

Let us look again at the vectors:

$$\textcircled{N_2} \quad N_1 \quad Y_1 - F$$
$$N_1 \quad \textcircled{N_3} \quad \textcircled{Y_3} - F$$
$$Y_3 \quad Y_2 \quad Y_2 - P$$

The third man will find it to his advantage to offer to change his vote on the middle issue if the first man will *not* trade off his vote on the first issue. Then the vote will go

$$Y_2 \quad N_1 \quad Y_1 - P$$
$$N_1 \quad Y_3 \quad \textcircled{N_3} - F$$
$$Y_3 \quad Y_2 \quad Y_2 - P$$

If we analyze the results of the trades in terms of payoff for each person (i.e., which of his choices win), we may display the outcomes as follows:

	P			F			P		
Issue A	P			F			P		
Issue B	P			F			F		
Issue C	P			P			P		
Voter I wins	2nd	3rd		1st	3rd		1st	2nd	3rd
Voter II wins	2nd	3rd		1st	2nd		2nd		
Voter III wins	1st	2nd	3rd	2nd			1st	2nd	

The last solution, $\begin{bmatrix} P \\ F \\ P \end{bmatrix}$, is stable and cannot be overturned by additional trades,[13] given the issues. The significance of the result is not, I emphasize, that it is the Pareto-optimal solution, but that it is the unique Pareto-admissible solution that results when each participant tries to achieve more of his own objectives by trading off things that interest him less. Pareto-admissible solutions in this mechanism are generated in the issue formation process, where there is incentive to draw the provisions of each issue as close as possible to its supporters' preferences so that it achieves as high a place on their ordinal ranking as is consistent with their underlying interest in the issue; and where there is a recognition that if it is drawn too strongly against the interests of those opposing it, the issue may be more vulnerable to defeat through vote trading.

In another article (see appendix B) I explored how the probability of vote trading increases as a function of the number of independent issues. Thus I suggested that a legislative body (the body set up to provide an orderly and efficient market for vote trading) should have a number of independent issues come before it for resolution. This happens in governments of general jurisdiction, but less so in special-purpose authorities, such as river basin commissions, where the issues are apt to be highly interrelated.

It may be, and I am currently exploring the possibility, that an environmental agency with authority over regional residuals management (air, water, and landfill) would have a sufficient number of separate issues (e.g., trade-offs between air quality and water quality) to enable vote trading to work. The exploration involves taking a region (the lower Delaware), constructing preference vectors for various districts in the region (municipalities, congressional districts, counties, equal-population districts, and so

---

[13] Under the assumption that an offer from voter II to voter III to return to $\begin{bmatrix} P \\ P \\ P \end{bmatrix}$, making both better off than they are at $\begin{bmatrix} P \\ F \\ P \end{bmatrix}$, will not be accepted as trustworthy.

forth) that reflect hypothesized preferences relative to air quality, water quality, landfill, tax rates, and the like, and using these vectors to estimate which quality standards and environmental policies are likely to be successful. (This will be discussed in detail in chapter 5.) One output of the process is to explore, as does the Dorfman model, whether changes in the constitution of the agency change probable outcomes.

Questions raised by this approach include:
1. What is the measure of the "best" outcome?
2. What criteria determine which political boundaries are used for decision making?
3. How reliable need the estimates of the preference vectors be?

ANALYSIS OF CASES

Perhaps the first statement in a critique about actual operating agencies, the Delaware Commission and the *Ruhrverband*, is to acknowledge that they work. Such an acknowledgment does not take us very far in an analytical way but is useful nonetheless, for I intend to keep a focus only on the connecting mechanism between preferences and social choices. This focus illuminates only a portion of an agency or a model and is not intended to provide light to judge the whole. With that caveat, let us move on to the questions raised in the first section.

Benefit Estimation

Both the Delaware Commission and the *Ruhrverband* use estimates of benefits to decide upon objectives and the means of achieving them. In each case the benefits to A are estimated by someone other than A. This neatly escapes the revealed-preferences problem in public goods provision, but does so by a father-knows-best principle. As a senior civil servant said to me recently in describing the Budget Bureau, "They know what you want and, by God, you better want it."

The estimation procedure in the Delaware case was a willingness-to-pay measure. The usual criticisms of this approach are that it neglects merit needs and income effects and that the estimation may be faulty. In order to get to what I consider the heart of the problem, I shall ignore those criticisms (any one of which may be telling in a particular case), and deal instead with a different issue, the appropriateness of willingness to pay as a benefit measure above a project level. The point is that so long as all publicly provided goods are not provided on the basis of specific willingness-to-pay estimates (as they cannot be), then the use of such a measure in one case must inevitably be faulty. What one man is willing to pay for water quality is surely related to what he already pays for other publicly provided goods, some of which he does not want. His political activity may consist of trying to shift a portion of his taxes to uses he prefers. Indeed, the whole debate on reordering of priorities is of this character. Analysis that assumes that the selection of a quality standard represents a marginal adjustment, other things changing only at the margin, cannot be squared with the reality of the interdependence of publicly provided goods. What such analysis does is to disguise the interdependencies, often with the active cooperation of politicians who do not want to be forced to make social choices. If the politician is willing to make the choice, then willingness-to-pay data *on one good* is of no help to him. In order to choose and construct his own preference vector, he must loosen all of the ceteris paribus assumptions this measure uses, solve in his mind all the issues simultaneously, and then see how he can improve his lot through vote trading.

Interest Representation

While the Dorfman model and my own work assume that "interests" get reflected as pressures on elected representatives,[14] in the Delaware and Ruhr cases "interests" are directly represented in the choice mechanism, either officially or unofficially. The advantage of such "representation" is clear. Technical options can be explored informally. Cooperation of affected parties can be achieved with a minimum of delay and misunderstanding. In the give and take of committee work, the possibility of workable compromise is enhanced. So long as such technical committees and "representation" is advisory, much good can be accomplished. When, however, the advisors become the policy makers (as in the Ruhr case), or when they are used to give the appearance of "public" participation in the decision process (as in the Delaware case), a different issue arises.

One must be clear on this issue, particularly since we depend, as a practical matter, a great deal on interest lobbying before representative bodies

[14] I gather that Dorfman is not particularly concerned whether the representative is elected or appointed. It is central to my work.

as a means of clarifying arguments and bringing out opposing points of view. The legislature, although it uses advocacy testimony to sharpen issues, is not a court, and it responds to voters. Replacing voter-elected representatives with representatives appointed by industry and/or municipalities interrupts the link between voters and their representatives. We tolerate the break in greater or lesser amounts, depending upon the politicization of the issues under review. At one time we were content to establish the Port of New York Authority as a way of achieving a fiscally solvent management. Now people in the area are in revolt against it because transportation and land use have become hot political issues.

Similarly, as environmental quality becomes, as it is becoming, a political issue,[15] it is less likely that interest representation will be allowed to dominate the decision process and less likely that appointed representatives will be allowed to make basic policy decisions about environmental quality.[16]

It is worth remarking, in passing, on the irony that it has been the misreading of political theory and the neglect of political history by the academic political scientist that has fostered interest representation in this country. In that misreading, Arthur Bentley's book on *The Process of Government* played a large role.

The Choice of a Voting Rule

The one-state/one-vote rule, used in traditional interstate compacts, and the vote that is proportional to financial contribution (the territory vote and the dollar vote) are both devices for weighting preferences and defining boundaries. Both are seriously deficient as means of translating preferences into social choices. In the one-state/one-vote case there are two distortions. The first occurs whenever the whole state is not affected by the decisions being taken (whenever the river basin covers less than the state). The second occurs because the people in populous states count for less than the people in less populous states. The legal background is quite clear and unexceptionable, but, as questions of environmental quality become more political, these distortions are apt to be under increasing attack.

The dollar vote is of obvious usefulness in corporate management and

---

[15] See, for example, the survey by Gladwin Hill, "Polluters Sit on Anti-Pollution Boards," *New York Times*, 7 December 1970, p. 1.

[16] In Germany, where traditional culture is not so attuned to representative government, the pressures for direct representation are probably not as great.

similar concerns where dollars are at risk. When lives, tastes, and wills are at issue, the dollar vote has little to recommend it. Moreover, unless the Wicksellian feature of voting simultaneously on program and budget is a uniform feature of the governmental sector, the dollar vote, as in the Ruhr case, remains an anomaly that works probably because there is no political issue at hand (largely homogeneous goals) or no tradition of self-selection of goals.

The Use of Political Weights for Analysis

A different kind of weighting is proposed in the Dorfman model. Dorfman does not advocate any weighted voting scheme. Although he is not explicit about whether he envisions a one-jurisdiction/one-vote scheme or a vote pattern that follows population, his discussion seems to favor population. He proposes, however, the use of a reasonable range of political weights for the actors (representatives on a commission, for example) for the analytical purpose of specifying the range of politically acceptable solutions. There are two objections to their use, only one of which Dorfman acknowledges. Dorfman recognizes that the choice of the set of reasonable weights is judgmental and that this creates a problem for the analyst. The more important objection, which Dorfman ignores, is that while one may take a decision and, by working backward, assign a set of weights that allows that decision, there is no justification for working through the process the other way.

A political process works because intensities of preference exist. These intensities allow trade-offs to occur. The result of these trade-offs determine the solution. The "weights" that produce the same solution exist, no doubt, but they exist in the same sense that Stimson's constant is a constant—it is always the number you need to get from where you are to the answer. The weights have no political significance at all.

There is even a way, for example, for a political process to work in the Bow Valley example constructed by Dorfman and Jacoby. It is difficult, because the commission really has only one decision to make—what water quality by what distribution of costs? Since the elements are interrelated, trade-offs are limited severely, but the decision can be cast in an issue formation framework.[17] In issue formation, trades are used to achieve a

---

[17] See the section on optimality considerations in appendix C.

more acceptable version of one issue—for example, by blocking versions neither party likes as well.

Preference vectors can be constructed for each of the four parties (cannery, Bowville, Plympton, and FWQA), and we may be as vague as Dorfman as to how these are reflected in the commission. Assuming they are reflected, and ignoring the possibility that there are more politically acceptable decisions than the fourteen generated by the weights used, we may construct an issue formation matrix as follows (excluding D-level solutions). First, rank the top three decision choices of each (indicated by subscripts):

Cannery	Bowville	Plympton	FWQA
$(12, 13, 7)_1$	$(4)_1$	$(14)_1$	$(10)_1$
$(6, 11, 14, 10)_2$	$(6)_2$	$(8, 9, 11, 13)_2$	$(11)_2$
$(9)_3$	$(7)_3$	$(10)_3$	$(12)_3$

Next, drop all decisions that do not appear in at least three columns. This is done on the plausible assumption that if anything is to be passed, it will have to be something a majority of the actors favor over the other alternatives. Rearrange to put the remaining decisions by rows (Y's are indicated, since we have assumed passage of something and are simply finding out which one):

	Cannery	Bowville	Plympton	FWQA
Decision 10	Y (ind.)	Y_1	Y_2	Y_1
Decision 11	Y (ind.)	Y_2	Y_1	Y_2

Decision 10 will be chosen unless trade-offs outside the issue of water quality occur. The indifference of the cannery is unlikely to be a source of bargaining strength, since none of the cannery's more preferred positions can win over 10 and 11.

This result does not depend upon the assignment of "political" weights to anybody, although the decisions were generated originally by the use of weights. The politically feasible solution space could have been calculated directly from the given cost and benefit functions, and preference vectors constructed within that space to define the trade-offs. The solution from that process would not necessarily be any of the fourteen decisions accepted by the Dorfman model. Whether or not the analyst has any business estimating political feasibility is another question and one which each man must answer for himself. He should know, however, that he is acting as a price-setter when he does so, and analysts who can correctly calculate equilibrium prices in actual markets are rare.

It may be useful, before leaving the discussion of political weights, to make a more general point. A critique of benefit-cost analysis made by Arthur Maass[18] in 1966 called attention to the political necessity for consideration of goals other than that of economic efficiency in public investments. His suggestion, however, was that the political process assigned weights to various goals and hence solved such investment decisions. It was a most unfortunate figure of speech, if that was what it was. If it was meant in a more precise sense, it was clearly wrong. As Koichi Mera, among others, has shown,[19] to solve any such trade-off between an efficiency and non-efficiency goal, both the total transformation curve (production possibility curve) and the social welfare function must simultaneously be known for an optimum trade-off to be specified. This may be represented by a diagram that plots percentage of efficiency gain as independent of some non-efficiency gain:

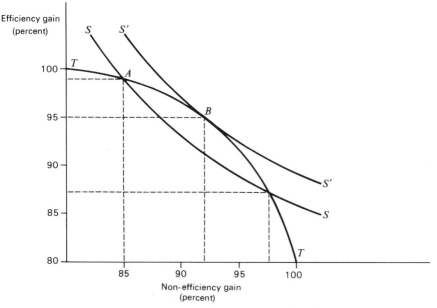

[18] Arthur Maass, "Benefit-Cost Analysis: Its Relevance to Public Investment Decisions," *Quarterly Journal of Economics* 80, no. 2 (May 1966): 208–266.
[19] Koichi Mera, "Income Distribution in Benefit-Cost Analysis," Discussion Paper no. 33 (Program in Regional and Urban Economics, Harvard University, May 1968).

Given TT as the transformation curve between efficiency gain and some non-efficiency gain, and SS and $S'S'$ as social welfare curves accompanying these two goals, then optimum point B (95 % efficiency gain and 92 % non-efficiency gain) can be identified. Any economic analysis that simply identifies the 100 percent efficiency point and, say, point A would not be providing sufficient information for any policy body or legislature to choose between the two. Even less would the trade-off ratio at A (99 over 85) tell the economic analyst anything about the shape of the social welfare function. Hence, it is as useless to ask a legislature to specify a single trade-off ratio (or constraint) as it is to ask the analyst to guess that point on the transformation curve which is relevant. If the analyst could supply the whole transformation curve, presumably the relevant policy-making body could supply the welfare curve that relates to it. That welfare curve cannot be inferred from past actions and will not result from a political consideration of *just* that transformation curve.

The criticism of using "political" weights is therefore not confined to the reasonableness of the weights chosen but refers also to the conceptual problem that goes back to Samuelson's statement about the need for simultaneously solving the income distribution questions and the public goods decisions. That is what an ideal political process does. Ex post one could calculate "weights" that would have given the same result, but the politician cannot specify those weights as he goes along. They "all depend" on all the others, in a very real sense. When one gets an "it all depends" answer from a politician, it is not always an evasion; it may be a literal translation of Samuelson's theory of public expenditures.

The Boundary Problem

Running through all four case discussions has been the unresolved problem of where to draw the boundaries of political jurisdictions over these social choices. What people should be included? When the country was new and politicians were more skilled in the conscious formation of social choice mechanisms, two principles were relevant to the drawing of political boundaries: they should be (1) large enough to encompass a heterogeneous mixture of people and problems so as to avoid the tyranny of the majority (Madison's defense of the American Republic against Montesquieu's criticism of large republics is based on this principle) and (2) small enough to ensure that most of the problems were of concern to most people. The latter principle was an attempt to guard against the possibility that many would have free money to make political deals—that is, that many issues would concern them not at all and they would be free to sell their vote to the highest bidder.

In modern times we find that the application of these two rules does not allow us to converge on any set of governmental institutions, and some economists in the Tiebout tradition can construct almost as many governments as people while others end up putting everything at the national level. In chapter 4, I will argue that the combination of technical interdependence of residuals plus these two rules makes a strong case for regional governments of general jurisdiction in many areas of the country. If that proposition is accepted, then districts within the overall jurisdiction must, on legal grounds and on utility grounds, be drawn in terms of equal population. Beyond that we have few guidelines. The size of districts will not, within some limits, make much difference except for two cases: (1) if the population is heterogeneous and the heterogeneity is, as is usual, reflected in the settlement pattern, then the districts must be small enough to reflect that, and (2) if the issues are likely to vary significantly from one area to another, then the districts must be small enough to allow those issues to be captured in the electoral process.

4

Residuals Management, Metropolitan Governance, and the Optimal Jurisdiction

(with ALLEN V. KNEESE)

The purity of the air we breathe, the quality of the water we drink, and the beauty of the landscape around us are functions, to some degree, of the kind of control we exercise over the residuals generated by our production and consumption activities. That such residuals are an inevitable result of production and consumption activities has been explicated in a recent book by Kneese and others.[1] The "management" of such residuals is, therefore, a continuing process that will be with us so long as we live in towns and cities from which residuals are sufficient in quantity and quality to overwhelm the natural assimilative capacity of the ambient air, water, and land.

It must be noted that estimates, mainly from federal sources, indicate that taking environmental quality seriously is going to cost us a lot of money. More specifically for the concerns of this paper, it is going to cost cities a lot of money. A recent compilation[2] of such estimates is that municipal ex-

This chapter is an amalgam of two papers. The first, "Residuals Management and Metropolitan Governance," appeared in Lowdon Wingo, ed., *Metropolitanization and Public Services* (Washington, D.C.: Resources for the Future, 1972); the second, "Environmental Quality and the Optimal Jurisdiction," was presented at the Joint Institute on Comparative Urban and Grants Economics, University of Windsor, 1972, and is used by permission.

[1] Allen V. Kneese, Robert U. Ayres, and Ralph C. d'Arge, *Economics and the Environment: A Materials Balance Approach* (Washington, D.C.: Resources for the Future, Inc., 1970).

[2] Jane Brashares, "Cost Estimates for Environmental Improvement Programs," appendix B of Allen V. Kneese, "The Economics of Environmental Pollution in the United States," in Kneese, Rolfe, and Harned, eds., *Managing the Environment: International Economic Cooperation for Pollution Control*, Praeger Special Studies (New York: Praeger Publishers for the Atlantic Council of the United States and the Battelle Memorial Institute, December 1971).

penditures between 1970 and 1975 to meet water quality standards alone may run to at least $14 billion and that an additional $34 billion would be needed if storm sewers were to be separated from sanitary sewers. Municipal expenditures for solid waste disposal, now running about $3.5 billion annually, are expected to rise by about $1 billion over the next few years. Expenditures needed in eighty-five cities to improve air quality by 5 to 15 percent with respect only to sulfur oxides and particulates are estimated to be at least $0.6 billion annually.

Estimates of the benefits of making such expenditures are much harder to come by because of estimation difficulties when dollar figures are appropriate and because of the many areas in which the political process is the direct arbiter of what the benefits are.[3] Nevertheless, some fairly hard data on the health costs of air pollution have been calculated. Lave and Seskin have estimated recently that a reduction of 50 percent in the air pollution levels in major urban areas would save over $2 billion annually in medical care and associated costs.[4]

Governments, of course, have other weapons in their arsenals than expenditures of money, and there is growing evidence that regulation of emissions, levying of user or effluent charges, and prohibitions of one kind or another will be used increasingly by municipal governments in their response to pressures for environmental improvement. A press report[5] of the proposed revision of the New York City Air Pollution Code lists the following items: phasing out the use of lead in gasoline, controlling nitrogen oxide emissions from power plants, limiting the volatility of gasoline, controlling the construction of new parking garages, limiting the sulfur contents of fuels, and outlawing the spraying of asbestos compounds. Other cities have either banned or are considering banning disposable bottles and washing compounds containing phosphates. The federal government has also passed stiff regulations on auto emissions and the use of lead in gasoline, and it is

[3] Aggregate estimates of costs and benefits, particularly those made by federal agencies that have a vested interest in one or another program, should not be taken as much more than indications of political strengths or weaknesses. As such, they indicate that federal bureaucrats and their politically sensitive masters feel that environmental improvement has some considerable potential as a source of programs, jobs, influence, and career making.

[4] Lester B. Lave and Eugene P. Seskin, "Air Pollution and Human Health," *Science* 169 (21 August 1970): 723–733.

[5] Quoted in *Environmental Science and Technology* 4 (November 1970): 882.

being urged to prohibit—or severely restrict—use of materials considered harmful to the environment.

All in all, there appears to be a tremendous concern about the residuals part of "residuals management" but not much concern about the "management" part. This is not said in a patronizing sense or to indicate a criticism of the administrative competence of federal, state, or local officials, but to point to two crucial facts. First, the materials-balance approach to residuals management, as developed by Kneese and others, emphasizes the interdependence of gaseous, liquid, and solid residuals. Suppressing one emission in one form in one place may well create a greater problem in another form in another place. If we are to manage residuals, therefore, we must take account of the physical interdependencies. Some of the interdependencies are commonplace and well known. For example, hauling garbage and sludge from cities to coastal waters for dumping exchanges one problem for another. It is easy to see why such practices appeal to coastal cities. The benefits of a "cheap" sink for solid waste are tangible, immediate, and desperately needed by the city, hard pressed to meet rising welfare and education costs with a shrinking or static tax base. The costs of using the ocean as a sink are as yet unknown, the incidence of the costs is diffused, and few individuals or groups have perceived a large enough impact to resist this use of the ocean.

The ocean, the air mantle, and many of the waterways of the nation are common-property resources, not "owned" by anybody. The land, on the other hand, is almost all in private hands (or under a specific governmental jurisdiction if in public hands). Thus it is that cities find it harder and harder to locate sites for solid waste disposal but did not until recently encounter much resistance to incineration or disposal of wastes in liquid form. Similar cases of mismanagement occur whenever one common-property resource (a river, for example) comes under public management while others (the oceans, for example) are still treated as free goods.

The second fact, therefore, has to do with the inconsistencies of governmental actions about residuals. These inconsistencies are of two types: local jurisdictions are free to do things harmful to other jurisdictions and, paradoxically, they are prevented (by higher governmental restrictions, prescriptions, and proscriptions) from doing other things that would be helpful. For example, federal subsidies to municipalities for sewage-treatment plants are very likely foreclosing the possibility of scale economies in treatment-plant construction and operation. Legislation that calls for percentage cutbacks on emissions from all plants is almost certain to be a costly way to improve air quality.

The prescriptive measures taken by state and federal agencies should be recognized for what they are—compromise measures on which a minimum political consensus could be achieved. Since they deal with a large area containing many watersheds, many airsheds, and more than one population cluster, it is almost impossible for such prescriptions to reflect the interdependencies spoken of earlier, or, indeed, any principles of environmental management. A crude national equity is about all that can be achieved—an equity that in fact costs everybody more money than it need cost.

A metropolitan area, particularly a standard metropolitan statistical area (SMSA) that includes a fairly large hinterland, could be considered an appropriate region for residuals management if some of the restrictions from higher governments were relaxed. Trade-offs among gaseous, liquid, and solid waste alternatives could be calculated. The assimilative capacities of the air, water, and land could be assessed in relation to the residuals load. Industrial, municipal, and individual decisions could be determined and, in large measure, controlled.[6] The political forms by which these assessments would be made and controls decided upon are, however, crucial, and there is at present vast confusion about political forms. The confusion manifests itself in practical arguments over city-county consolidation, metro government, special districts, local independence, and the like, but the confusion is, in reality, a confusion in theory.

RESIDUALS MANAGEMENT AND SOCIAL CHOICE

The three main political (social choice) problems that occur in residuals management are:

1. The choice of what *levels* of environmental quality the area is to have involves setting the level(s) of air quality with regard to, say, SO_2, particulates, and CO, setting levels of water quality in terms of allowable heat and BOD loadings, and decisions about aesthetic aspects of solid waste han-

[6] Analytical models to serve these purposes are under rapid development. See the paper by Clifford S. Russell and Walter O. Spofford, Jr., in Allen V. Kneese and Blair T. Bower, eds., *Environmental Quality Analysis: Theory and Method in the Social Sciences* (Baltimore: The Johns Hopkins University Press for Resources for the Future, 1972).

dling (dumps, for example), which raises issues of land use. Clearly these issues present a collective choice problem because, as we have pointed out, they involve common-property resources and have public goods characteristics. Furthermore, even if preferences could be perfectly known, no optimal levels could be unambiguously specified except in highly exceptional cases.

2. Choices must be made about the income *distribution* issue as it relates both to benefits and to costs. There will be several ways of achieving any specified level of quality. Each may have a different distribution of benefits and costs.[7] Urban economists have generally been content to assume a reshuffling of people (voting with the feet) as a way in which individuals try to choose the best available mix of public services, including environmental amenities, within their budget constraints. Such movements, while they occur regularly, are of limited help, for several reasons. The collective choice issue cannot be avoided by ambulatory elections.

3. Management *instruments* must be chosen collectively. Although this choice is closely connected to the distributional issue, it is a distinct problem, for it also includes the efficiency issue. In residuals management, one can achieve a given ambient concentration of, say, SO_2 either by setting emission controls (each source restricted to some quantity of SO_2 emission, not necessarily the same for all) or by setting effluent charges on each source. Clearly each course will have different distributional effects. But there will also be efficiency differences in the aggregate between the two methods of control. As we have indicated, opting for effluent charges may well bring large efficiency gains for the overall residuals management systems. Moreover, there is the matter of providing collective facilities, such as reservoir systems, reaeration equipment, and regional treatment facilities.

The essence of a political entity is that it possess a means of making collective (social) choices. This is true regardless of which choices are left to individual market decisions (unless all of them are) and which are deemed collective. So long as some important decisions are not made on an individual basis, then some machinery must be set up to make them collec-

---

[7] Indeed, in models prepared at RFF the variations of both costs and benefits are wide even over very small changes in levels of ambient concentrations. See Clifford S. Russell, Walter O. Spofford, Jr., and Edwin T. Haefele, "Environmental Quality Management in Metropolitan Areas" (Paper presented at the International Economics Association Meeting, Copenhagen, 19–24 June 1972). Publication forthcoming.

tively. Air quality is a good example of a decision that must be made collectively if it is made at all. A number of problems emerge as air quality becomes a social choice issue, and they illustrate the theoretical issues involved. They may be posed as follows:

1. How large an area is to be affected by the decision on air quality?
2. What air quality level(s) should be established?
3. How should the expense of attaining these levels be borne?
4. By whom should the decision be made?
5. By what means should the air quality levels be achieved?

The answers to these questions may be constrained by technological limits (production possibilities), by economic limits, and, in practice, by political and financial limitations.[8] Although these limitations will dominate the problem in most specific situations, underlying them is a more fundamental issue. What theory or normative model could be used to guide our answers?

The last question addresses the issues of "participatory" democracy, economic efficiency, the responsiveness of the system, minority representation and control, income redistribution, community values in planning, and the other value conflicts about goals and methods. The questions are especially hard to discuss because of the wide disparity of views, most held subconsciously, about them. Nevertheless, it is essential to make the attempt.

In an attempt to constrict the area of disagreement, let us assume that everyone would agree that decisions cannot be made in the absence of information about the following:

1. present air quality at various locations in terms of sulfur oxides, particulates, and so forth;
2. what residuals are now being discharged into the atmosphere at what points;
3. alternative methods of achieving reductions of ambient concentrations and the costs (including employment effects) of such reductions.

There will be some anti-system people unwilling to allow even that degree of information collecting, but ignoring them, we have already established a

---

[8] They are, most certainly, influenced by the trade-offs among residuals—i.e., higher air quality levels may be achieved at the expense of lower water quality.

need both for a monitoring function and for some technical analysis of alternative actions for improving air quality.

If we now admit the interdependence of physical forms of residuals, we cannot resist monitoring and analyzing alternative costs and impacts on waterborne wastes and solid wastes. In other words, regardless of how we decide to handle the questions of quality goals, extent of coverage, financing, and the machinery for choosing *for air*, we have committed ourselves to fairly elaborate residuals monitoring and analysis for the air, water, and solid waste of some region. If we are to be rational, there is no escaping the obligation to perform these tasks.

Keep in mind, however, that we have not committed ourselves to any particular way of performing them. We may have a civil service perform them; we may contract with a private company; we may create a public corporation. All three are in fact done.

We now are contemplating a metropolitan (at least) monitoring and analysis system for gaseous, liquid, and solid wastes. We have arrived here by taking the obvious and least controversial options. We are not far from the view of most experts in the field. Still, in our attitude of suspended disbelief, let us ask this monitoring and analysis system to perform on a metropolitan (at least) basis and look at the results. The results could, of course, be transmitted to the state and/or to each local jurisdiction in the area. No specific governmental form is necessary for this much of our system to work, just a little information cost-sharing. It is worth calling attention to this fact, because many assume that a metropolitan government, or metropolitan special authority or district, "must" be set up in order to get this information. Moreover, it is even possible that voluntary, unanimous agreement by all municipalities in the area could result in the establishment of a program based on information gathered through such a monitoring and analysis system and that enforcement could similarly be agreed upon and undertaken, even on a contract basis. Councils of government can exist and function by this Wicksellian machinery.

If we wish to move beyond the strictures of the unanimity rule and councils of government, we face some serious theoretical and practical questions, the answers to which are far from clear. Our monitoring and analysis service may tell us, in some detail, how we may efficiently raise air quality (in a total residuals management context) by various increments. The impacts and benefits of such moves will probably vary over the territory involved.

How we move on the decision now makes an enormous difference. If we ask municipalities, the response will probably be different from the response of county governments. If we ask interest groups, the response might be still different. If we imagine equal-population wards, the answer is again different. Moreover, a referendum vote will produce a different answer from that worked out in a legislative process of representatives in the same jurisdiction. A survey, conducted for a planning commission, might produce another answer. A hearing would generate its own unique confusion.

It is important to stress that the answers would not be different simply because of inefficient information-gathering processes. The answers are different because the way the questions would be asked, the way the issues would be formed, and the process of resolution or decision by the different units would all be different.

Whom do you ask? What proposal(s) do you ask about? Framing the issues for decision and deciding upon the appropriate collectivity to ask is the essence of the government problem. It is at the heart of the value conflict problem. It is at the heart of the welfare economics problem with a social welfare function. It is the nexus of confusion in much social science theory.

An attempt has been made elsewhere (see appendix C) to provide a way out of the thicket through a utility analysis. It showed that individual values could be "appropriately" related to social choices through a two-party system of representative government. The criterion for "appropriateness" was that the government choose the same policy as would be chosen if everyone were in an assembly and vote trading on the issues were allowed. Representative government, under a two-party system competing for votes in single-member districts, proved to be a mechanism capable of producing solutions identical to those chosen by direct voter trades.[9]

The issue of how to choose the collectivity, that is to say, what boundaries to use, is yet to be faced. Madison suggested that the rule for boundary setting for governments of general jurisdiction was to encompass a heterogeneous population with common problems. Applying that rule still leaves us with a large number of possible choices. The physical interdependencies in residuals management, referred to previously, help to narrow the choices.

---

[9] This statement should not be taken to be an assertion that present, existing governmental forms or parties are capable of producing this "ideal" outcome any more than present businesses sell at competitive equilibrium prices.

For example, the economic reach of a metropolitan area, its economic base, is reasonably consonant with residuals management problems associated with air quality and solid waste. Moreover, the economic reach is a useful one for considering a tax base, a transportation system, and land-use controls.

An entire watershed is, however, rarely encompassed by a metropolitan region, and water quality management is most efficiently done on a watershed basis. Thus, problems of coordinating executive government and of setting legislative policy exist because of this incongruence of problemshed boundaries. The usual approach to watershed management, the interstate compact, has its own set of problems (see chapter 2) but may be a useful solution if state lines are reasonably consonant with the watershed. When they are not, a separate executive agency for watershed management, controlled by a board composed of representatives from the concerned metropolitan regions, may be the most easily arranged way to effect policy coordination while realizing the efficiency gains possible from basinwide management of water resources.

We suggest that a new government is most apt to be successful at the metropolitan-regional level (where a number of common problems appear to coexist and the population is heterogeneous) and that several options exist to make the necessary ties between that general government and the watershed. We are led to this conclusion in part because the legislative mechanism for making social choices—vote trading—can function more efficiently when many independent issues (transportation, land use, tax rates, air quality) arise in the same territory (see appendix B), and in part because the residuals management (executive mechanism) function can be carried on efficiently at this level.

It may be worth remarking that existing states might qualify on most of the points made. Are we just re-inventing state government? That we are not is emphasized by a recent action of the Maryland legislature regarding a Washington Metro (transit) bill.[10] Regarded as a "local" issue (not a common problem), the vote trading necessary on the part of the Washington area representatives was far beyond that which would have been required in a metropolitan legislature. The reason: since most state legislators had

---

[10] Reported by the *Washington Post*, 10 April 1971, under the heading, "The Metro Bill and the Sea Full of Sharks."

no interest one way or the other, intensities of preferences, which are the essence of vote trading, played no part. In effect, free money was introduced into the system, driving up prices for all who *were* interested in the outcome. On a metropolitan level (where Metro is a common problem), all representatives would have *some* interest in the transit bill, and the intensity of those interests could be appropriately expressed through vote trading.

It is also worth noting that whereas a legislature composed of, say, equal-population districts within a metropolitan region provides a means of aggregating individual preferences, *given the issues*, it is quite likely to be parochial as a generator of issues. Thus some provision should be made either for some representation at large in the proposed legislature or for an elected executive with veto power. The former ensures that metropolitan interests get introduced into the deliberations; the latter goes further and ensures that these interests must concur on any action undertaken.

Collective Choices and Common-Property Resources

The rudimentary and ghostly forms of executive and legislative bodies we have hypothesized are sufficient to confront a major practical problem. Again, although it is practical, the reason it is a problem is that there is some confusion in the theory.

The quickest way to get into the problem is to say that the analysis section (executive) of our hypothesized government will be perfectly willing to make policy de facto, using its own estimate of political reality, and the legislators will be tickled to death to let it do so, under the pretense that it is pursuing an objective, scientific method of decision making by which costs and benefits are appropriately measured, citizen and special group interests "taken into account," and the final choice (albeit clearly signposted) actually "made" by the legislature. Since most of the readers of this paper will recognize this situation and have their own reactions to it (ranging from dismay to joyful acceptance), we shall not dwell on a description of a situation that has put most of the real choices in executive hands, relegated the legislature to a legitimating role, and allowed many legislators to be all things to all people.

Our argument here is that, failing a return of royal prerogative as the basis, there is no way the executive branch can make the collective choices consonant with the theory of representative government and no way that collective choices can be avoided in decisions about resources that are common property.

The first point can be disposed of quickly. Even though a mayor, governor, or other executive official is elected, he is elected by *one* majority. It is almost a truism of representative government in America that the only defense against the tyranny of the majority is to lodge policy determination in the legislature, where different issues bring forth different majorities and intensities of preferences can be used (vote trading) to arrive at final decisions. The only analogue to this process in the executive is the election strategy by which a candidate constructs a bundle of issues and positions to run on. One could imagine the two processes coming to the same conclusion, but it requires the imagination of a socialist planner who says he can come to the same equilibrium solution as a free market system does. It is an interesting, though pathetic, commentary on our present predicament that cyberneticists[11] now try to replace representative government by a system of feedback loops from executive policies to citizen reactions to new executive policies. This approach, which rejects the distinction between legislative and executive functions, puts us back to the organic state of Tudor times and ignores the constitutional battles of the seventeenth century. But, since American constitutions, federal and state, are the direct results of those seventeenth-century battles, such approaches create a paralysis of government. The paralysis occurs because of attempts to force executive governments to do what constitutional machinery prevents them doing—to make collective choices. The constitutional bar, a result of political history, is valid on technical grounds—executive government choices fail to converge (the historical demonstration occurs convincingly in the Protectorate), and they are not stable (since trade-offs can be neither discovered nor efficiently carried out except in the most coarse-grained sense. The reign of James II illustrates this failure.)

The second point, the necessity for collective choices about common-property resources, must be raised if only because of attempts by some to force all such questions back to marketplace or pseudo-marketplace solutions. The incentive to do so is, of course, great. Economic theory suggests that individual decision making about what kinds and amounts of goods and services to buy will lead, through a market exchange system, to the

[11] See E. S. Savas, "Cybernetics in City Hall," *Science* 168: 1066–71.

optimal level of production of these goods and services, given certain assumptions about ideal markets, and lack of externalities (third-party effects). Such theory proceeds from a given distribution of income and assumes that all goods and services can be divided at the individual level—i.e., are not consumed in common.

For goods and services not capable of being so treated, a more complicated theory is needed. Specifically, we must consider distributional criteria as well as efficiency criteria. The income distribution problem associated with these "public goods" is particularly difficult because the consumption of the good cannot be differentiated among consumers on the basis of their voluntary choice in markets. When the supply of a public good changes, both efficiency and distribution are affected, and there is, in general, no way to be sure that equating marginal cost with the sum of marginal willingness to pay will be a welfare maximum. This has been a hard problem for applied public economics, and several devices have been used to get around it. Otto Eckstein, in his work on water development,[12] explicitly assumed that the marginal utility of income (a cardinal measure not employed in contemporary analysis) is the same for all individuals. This effectively wipes out distributional considerations, but most economists regard it as grossly unrealistic.[13]

The assumption most often made implicitly by practicing economists is that it is a mistake to consider individual public goods as situations in isolation. Rather the whole complex of public goods should be considered. Some will affect one group adversely and another favorably and others vice versa. Thus there will be a lot of cancellation of distribution effects. The society that makes its decisions based on efficiency criteria will be one in which most people will finally be better off than one in which criteria are used that foreclose efficient solutions. It is further assumed that public goods are a

[12] Otto Eckstein, *Water Resource Development: The Economics of Project Evaluation* (Cambridge: Harvard University Press, 1961).

[13] Another device is to assume $\delta s^i / \delta y^i$ (where δs^i is the marginal rate of substitution of private goods for the public good of individual i, and δy^i the marginal change in income of individual i) is the same for all individuals i, then one can separate the distribution problem from the allocation problem. That is, different distributions will not affect the optimal allocation. In applied ordinal models this assumption is necessary for consistent aggregations of consumer surpluses. Karl-Göran Mäler has proved this point in *Environmental Economics: A Theoretical Inquiry*, to be published by Resources for the Future, Inc.

rather small part of the economy and that private goods are allocated (through tax and subsidy policy) in an ethically sanctioned way.

A second, but separate, problem has to do with the measurement of willingness to pay. Samuelson[14] has argued that the willingness to pay for public goods cannot be measured because exclusion is not possible. This is a frequently repeated fallacy in the literature. It results from the fact that most theoretical discussions start from the idea that one would ask people what they would be willing to pay. In practice this technique is hardly ever used. It is also not correct to say that persons cannot be excluded from the provision of public goods. Frequently exclusion is *possible*; but if we are talking about a pure public good, it is *undesirable* on efficiency grounds. Exclusion arguments have done much to muddy the waters. There is a continuum of exclusion problems with most public goods near one end and most private goods near the other. It is nevertheless true that markets do not provide *direct* information on the value of public goods and, therefore, estimation problems are always difficult and sometimes close to impossible. Accordingly, even when strong efforts are made to obtain willingness-to-pay information on the range of public good candidates, it is likely to be partial and of widely differing dependability from case to case.

There has been much work on willingness to pay in the context of common-property resources, primarily because the advantage of using the price route is that the allocation is then automatic and need not take anyone's time (beyond a policing or enforcement capability). The advantage is significant. A further advantage is that revenues are generated, and no better source of revenues could be imagined than charges on effluents. A third advantage of prices is that they can be efficient—encouraging firms, municipalities, and individuals to seek lower-cost methods of dealing with their residuals through changes in their amount, timing, and physical forms.

The mechanism of price, as such, is unexceptionable. The level of prices is another matter. We can easily agree that relative prices should depend on relative damage caused to the resources or to the relative costs to process the effluent. The *level* of prices, on the other hand, relates to a quality specification. What quality air shall we have? Can we find *that* out by estimating people's willingness to pay? Let us explore this idea under the best possible assumptions, ignoring the distribution-of-income problem, ignoring the

[14] "The Pure Theory of Public Expenditures," *The Collected Scientific Papers of Paul A. Samuelson*, ed. Joseph E. Stiglitz (Cambridge, Mass.: M.I.T. Press, 1966), p. 1223.

estimation problem, and assuming that the externalities issue is incorporated in the pricing schedule and thus can also be ignored. In other words, if income redistribution is set aside, if we possessed schedules of people's willingness to pay (setting aside the estimation problem), and if our pricing scheme accounted for all external effects of each person's decisions, could we determine the appropriate level of air quality directly by measuring willingness to pay without asking for a collective choice on the matter?

Clearly, under this highly artificial set of assumptions, we could equate marginal cost to the sum of marginal willingness to pay. We could, therefore, provide the "optimal" level of air quality, or that level beyond which no one is willing to bear the next increment of cost. But, since there are physical links between air, water, and solid residuals, we should have to expand our willingness-to-pay schedule to encompass liquid and solid as well as gaseous residuals in order to arrive at environmental willingness-to-pay schedules. Moreover, since environmental quality is only one of many publicly provided goods and services, we cannot ignore the interrelationships with these other goods and services.

In short, a partial solution based on willingness to pay will not suffice in an equilibrium setting. We should have to have willingness-to-pay schedules on all publicly provided goods. It may be wholly appropriate to use willingness-to-pay data in a recreation project analysis as a guide to investment in one park, lake, or whatever. This is a legitimate partial analysis in which executive government, under some policy authorization and budget constraint, is attempting to discover (much as a private firm would) where the payoff is highest. It is a totally different thing to use a willingness-to-pay measure to ascertain how many parks or lakes an area should have (or what level of air quality). The latter questions are better asked by political entrepreneurs who are responsible for general taxation if intensities of preferences about air quality are to be used to accommodate policies or investments in other areas where there are intense preferences also. Assessing willingness to pay may, however, play a very useful role in helping to focus the political process, particularly when it can be shown that large welfare gains are possible. Such evidence gives incentive to political entrepreneurs and ammunition to the public. The resulting collective choice, which provides the simultaneous solution to the efficiency and income questions, benefits from both. Indeed, it may be that estimating willingness to pay and providing other measures of benefit and cost (including incidence of such benefits and costs) are especially useful as issue formation devices for the process of collective choice.

THE RECEIVED THEORY OF OPTIMAL JURISDICTION

We are left, however, with the question of who should be included in what collectivities. For several years a few economists have studied the question of optimal jurisdiction for the provision of public goods. We do not wish to go as far as H. L. Mencken, who is said to have remarked, "For every problem economists have an answer—simple, neat, and wrong." But the results of this research do appear to be leading down a rather narrow path.

A classification of criteria suggested by Rothenberg provides a useful starting point for our discussion of this literature. He suggests four (which often conflict):

1. Minimize political externalities.
2. Minimize crossjurisdictional externalities.
3. Minimize the resource cost of providing public goods.
4. Maximize achievement of redistributive goals.[15]

The last three are straightforward in concept. Items 2 and 3 will be considered in some detail in the last section. The last point, *as a general matter*, seems to us largely inapplicable to our topic, which surrounds the regional scale and structure of government. As others have argued, this scale of government is not suitable for undertaking *general* redistributions of income.[16]

But we hasten to point out that this conclusion does not extend to redistributions of income inherent in the provision of public goods to nonhomogeneous populations. Since a public good is supplied in bulk, so to speak, to a large number of persons, and the amount the individual gets is not under his individual control, redistribution occurs simultaneously and inevitably with the provision of the good unless canceling lump-sum trans-

---

[15] J. Rothenberg, "Local Decentralization and the Theory of Optimal Government," in J. Margolis, ed., *The Analysis of Public Output* (New York: Columbia University Press for the National Bureau of Economic Research, 1970).

[16] We must note, however, that to the extent that general income distribution is regarded as a problem for this level of government, its solution requires nonhomogeneous jurisdictions. For a discussion of this issue, see Paul V. Pauly, "Income Redistribution as a Local Public Good" (Paper given at Public Choice Society Annual Meeting, Pittsburgh, 1972).

fers (or equivalent tax arrangements) can be made. The concept of lump-sum transfers has been an immensely useful device in the elaboration of public goods theory, but it is entirely inapplicable in practical situations. Accordingly, criteria based strictly on efficiency tend to be unacceptable, and the redistribution issue becomes a central one that the political process must somehow resolve. We examine the functioning of a representative government process in this connection in some detail in the next section.

Here we wish to come to a focus on what Rothenberg, in his listing of criteria, has termed "political externalities." Consideration of this matter is important, since the "voting with the feet" theory, so prominent in the recent literature, is focused mainly on it. Most simply, political externalities means imposing the majority will on minorities: if 51 percent of the people vote for something in a majority-wins election, 49 percent are not represented, no matter how unhappy they may be with the outcome. For reasons we develop shortly, we believe this argument betrays either a profound misunderstanding of the representative government process or an unwarranted emphasis upon an issue-by-issue referendum approach to collective choice. Of course, if one believes that the concept of political externalities provides insight into the question of optimum jurisdictions, then (aside from conflicts with the other criteria) the clear implication of the argument is that people with homogeneous preferences should be grouped into separate jurisdictions.

Before criticizing this argument further, it may be useful to present it a bit more fully. The "voting with the feet" viewpoint stems back to a seminal article published more than fifteen years ago by Charles M. Tiebout.[17] He assumed mobile consumer-voters and flexibly redefinable local jurisdictions which can be independent of each other in the sense that there are no inter-jurisdictional externalities. He concluded, among other things, that for a given configuration of jurisdictions (group size, collective output, tax rates, and expenditures), each consumer-voter "moves to that community whose local government best satisfies his set of preferences" and, further, that "the greater the number of communities and the greater the variance among them, the closer the consumer will come to fully realizing his preference position."[18]

[17] Charles M. Tiebout, "A Pure Theory of Local Expenditures," *Journal of Political Economy* 64 (October 1956): 416–424.
[18] Ibid., p. 418.

A recent paper by McGuire,[19] by far the best of this genre, offers a proof of this proposition under the following assumptions:

1. Population heterogeneity is taken to mean that there are multiple subpopulations, each homogeneous unto itself and discretely (in the mathematical sense) separated from every other homogeneous subpopulation—every member of a homogeneous subpopulation has the same demand curve for the public good.
2. Each homogenous subpopulation is large enough to form many efficient jurisdictions in the sense of being able to provide any particular public good at issue at minimum average cost.
3. A separate and distinct jurisdiction can be formed for each local public good.

The implication of these assumptions is that "it is [aside from transactions costs] Pareto optimal to allow homogenous (with respect to marginal rates of substitution) groups to form their own collectives." This conclusion is derived by means of an elegant mathematical analysis. We will not repeat it here, because the central reason for the result is intuitively very plausible. We should note that the problems of crossjurisdictional externalities and minimizing the cost of public service either are ruled out by or very straightforwardly follow from the assumptions. No political externalities (majority effects on minorities) are imposed: the result would be determined by a unanimous vote if the issue were the amount of public good provided, given the equal cost-sharing scheme. A compulsory cost-sharing scheme must be specified to avoid the so-called free-rider problem. Since the jointness-in-supply assumption means that no one in the relevant area can be excluded from receiving the public good, its provision cannot be arranged via voluntary individual payments.

McGuire's particularly clear and precise analysis provides an excellent frame for further discussion of the "voting with the feet" solution to the jurisdictional problem with respect to the public good and externality aspects of residuals management problems. To this we now turn.

A Critique of the Received Theory

The severity of the strict homogeneity assumption is, perhaps, self-evident, but not sufficiently so to have kept a generation of urban econ-

[19] Martin McGuire, "Group Segregation and Optimal Jurisdiction" (unpublished).

omists from relating Tiebout's theory to the widely observed phenomenon of persons moving around in a metropolitan area, and then drawing the conclusion that this results in homogeneous jurisdictions. Clearly there is more homogeneity than there would be without such moves, but far too little to give the Tieboutian models much real-world significance. Simply reflect on (1) the history of self-selected communes where homogeneity of preferences is vastly greater than in most other communities, and (2) the last PTA meeting you attended in your homogeneous suburb.

That we do not regard voting with the feet as a general solution to the public goods problem does not mean that we fail to recognize a role for it. It is useful to think of two categories of public goods (bads) situations. In the first are those which can be obtained (avoided) only by locating in par-ticular places. For example, if one feels strongly about either living with or not living with a certain ethnic group (we pass no moral judgments here), there is no way of meeting one's desire except by location decisions. The same is true of access to landscape or other fixed locational aspects of an area. Of course, moves to meet preferences of this type will not fully solve the public goods problems, since, contrary to McGuire's simplifying as-sumption, multiple public goods are always supplied by every jurisdiction or by multiple jurisdictions overlapping the same space. Nevertheless, if a person feels strongly about some fixed locational feature of the area, he may well move and thereby improve his welfare, even though this does not permit him to enjoy the precise mix of public goods which he desires. That is to say, he might prefer more or less of some public good, given the cost of providing it and the share he would bear even under a nondistributive, cost-bearing scheme.[20]

The other class of public goods consists of those, the supply of which can be altered, at thinkable cost, by collective action. There are many such—quantity and quality of education, transportation facilities, and police and fire protection are all examples as is, of course, the quality of environmental media as affected by residuals generation and disposal practices. Public investment, regulations, and prices can and do alter all of these. In these instances, locational choice is not the only adjustment and, in view of the multiple public good characteristic of each location, often is not an efficient

[20] For discussion of the nondistributive, so-called Lindahl solution, see Martin C. McGuire and Henry Aaron, "Efficiency and Equity in the Optimal Supply of a Public Good," *Review of Economics and Statistics* 51 (February 1969): 31–39.

adjustment. This is not to say that, in the absence of government institutions of appropriate scale and capability for making collective choices, it may not be second best. It is at least arguable that mobility has been a substitute for, and perhaps even hindered the development of, political institutions that could have been instrumental in attaining greater aggregate welfare. Private costs (commuting, for example) may have been substituted for public goods (better air quality in central cities) in part because no appropriate institutional vehicle for collective action was available. This suggests an important role for higher units of government in organizing institutions for collective action at the local and regional level. Indeed, we regard the question of political institution building through public policies at the federal level as one of the most important and least recognized problems facing the national government; we will return to this matter in the final section.

Within the geographical scope of the large "problemshed" authorities required for efficient environmental management, however, numerous, quite diverse public goods will have to be provided, and the population simply cannot be very homogeneous within such spaces.

The more fundamental point, however, is that homogeneity is not needed to deal with political externalities. To oversimplify somewhat, the *raison d'être* of any body politic *is* to handle political externalities. Were any group of people totally homogeneous, then the only reason for government would be to avoid free riders. Any one individual could be designated to make the choices, since all choices would be the same. Government is necessary in part because no two men (not to speak of large groups) ever agree completely about public goods. In democratic regimes, where individual preferences form the basis for social choices, the heterogeneity of the community is the engine that resolves social conflicts, that is to say, makes social choices.

A famous case may illuminate this assertion. In the first Congress following the adoption of the Constitution there were two large issues, neither of which could at first be resolved. One was the question whether the United States would assume the wartime debts of the states—the South said no, the North yes. The other issue was where to locate the U.S. capital—the South said on the Potomac, the North said no. Hamilton was in despair. However, when Jefferson returned from Paris he called the leaders of the two sides together and uncovered an additional fact that had escaped

Hamilton's attention. He found how each side ranked the two issues. With subscripts for ranking, the situation was as follows:

	North	South
Potomac site	N_2	Y_1
Assume debts	Y_1	N_2

He quickly saw the possibilities for a political transaction. Accordingly, he arranged a vote trade, with the North agreeing to the Potomac site in exchange for the South's agreeing to assume the states' debts.

Any representative assembly deals similarly with intense minority opinion, so well in fact that many complain of the undue strength of minorities under our system. It is worth noting that what usually passes for homogeneity in "voting with the feet" arguments is in fact agreement of ordinal ranking of issues, for example, that "good schools are most important." What happens in this situation is that people disagree on the means to achieve them, resulting in a vote as follows:

	Voters				
	1	2	3	4	5
Specific school issue A	Y_1	N_1	Y_1	Y_1	N_1

No trades are possible on this issue, since everyone ranks it first. The initial majority prevails and the minority suffers without any compensating gain. Thus homogeneity in an ordinal sense may well bring on a majority tyranny.[21]

Although vote trading permits dealing with heterogeneity in a single political jurisdiction, it does not directly address the question of appropriate jurisdictional boundaries. Madison's implied definition for a political jurisdiction was a heterogeneous population having common problems. This probably had much to do with giving us territorial representation, as opposed to interest representation, but it says little about how to carve up the territory.

THE GENERAL-PURPOSE REPRESENTATIVE

It is possible that the boundary problem can be finessed by electing a general-purpose representative (GPR) at a district level (the districts being

[21] The theory underlying the preceding points is elaborated in appendix B.

smaller than any general government). Such representatives would sit in all local and regional governmental bodies having jurisdiction over the district (as measured, say, by taxing authority or control over land use). The purpose of this building-block approach would be twofold: (1) to give the representative control over the whole range of local issues so that he could use his vote in one assembly as a lever in another, thus providing an opportunity for registering intensity of preferences, and (2) to enable governments of varying territorial reach to be assembled (and perhaps more important, disassembled) easily and conveniently with no damage to the basic political fabric of the area.[22] The suggestion has the added advantage of focusing local politics so that citizen participation in the party structure and electioneering have more potential payoff, since the representative is less vulnerable to special-interest groups if he sits on the sewer board, zoning board, and the school board, as well as on a local government council, than he would be if he sat on only one board. It seems reasonable to assume that the GPR system could go a long way toward overcoming voter apathy in local elections, since all issues would be focused in the election of one man, much as they are in the election of a governor, a senator or representative, or the president. Moreover, by focusing all issues in one election, the tendency for majority tyranny to emerge (in small populations such as the district) is mitigated. Intense preferences of minorities on one issue can be used to advantage in electoral politics[23] as well as in legislative politics.

Moreover, in the context of our present concern with environmental management in a metropolitan area or region, the general-purpose representative provides, perhaps, a way out of the dilemma posed by the mismatch of jurisdictions to problems. If we have a management agency whose reach encompasses the problem (an expanded river basin commission, for example), we have only to put such an agency under the policy control of a representative body composed of the GPRs covering that region. Although such a move may appear to make the agency answerable to too many bosses, in fact it gives the agency clear policy direction from one source.

[22] For example, a district could opt into any local government body by being willing to be taxed by that body. It could not opt out without the permission of the governmental body concerned.

[23] For a clear presentation of the case, see John E. Jackson, "Intensities, Preferences, and Electoral Politics," Working Paper 705-72 (Washington, D.C.: The Urban Institute, March 1972).

It will be apparent that *how* the district lines are drawn will influence the outcome of any preference aggregation procedure.[24] Equal-population districts will be necessary to meet Supreme Court tests, and the gerrymander is an ever-present event in the drawing of any political line. This question will, however, be begged, as it is a universal one and not unique to local districts. Moreover, the present political party structure provides an adversary procedure to cope with the problem.

In a recent paper,[25] Orbell and Uno provide some empirical evidence of who leaves an area and who stays and why. In brief, they suggest that among whites, higher-status people are more apt to stay and fight and lower-status whites are more apt to move when neighborhood problems arise. Blacks of similar status are more apt to stay than whites, having fewer opportunities to move. For whites, there is a tendency to stay, once a move has been made, for three to six years, during which political action efforts are greatest. Beyond six years, moving increases again as a political solution. Among blacks, political action increases monotonically with length of residence.

Such evidence, scanty though it may be, indicates that people do try to change their neighborhoods in ways other than by moving. That fact, together with recent Supreme Court rulings granting instant voting privileges (no more year-long wait to participate in local elections) and other Court rulings demanding equal public services in all areas of any political jurisdiction (if one street in Suburb *A* gets street lighting, then all streets in Suburb *A* should get street lighting) leads us to speculate that general-purpose governments will be getting smaller and their representatives will be elected more frequently.

Not a shred of empirical evidence exists to support the speculation. We simply find it unlikely that equal services can be supported across large jurisdictions, and hence feel that smaller jurisdictions will provide a way out of that problem. Similarly, if local pressures for participatory democracy grow, we may find ourselves back to the days of the early Republic, when annual elections were considered *de rigueur*, that is, when semiannual ones were not required.

[24] See Russell, Spofford, and Haefele, "Environmental Quality Management," pp. 55–58.

[25] John M. Orbell and Toru Uno, "A Theory of Neighborhood Problem-Solving: Political Action vs. Residential Mobility," *American Political Science Review* 66 (June 1972): 471–489.

Both suggestions run counter to the orthodoxy of consolidated, large governments required on efficiency grounds and long intervals between elections sought for "stability" reasons, so that long-range planning can be effected. We suggest that the orthodoxy confuses the needs of executive management with those of representative government. Were we to couple our district GPRs (frequently elected) with large area management agencies (metropolitan transport commissions, environmental management agencies, planning commissions), then it might be possible to achieve efficiency in planning and execution with responsiveness to shifts in population. Since the preceding statement is anything but intuitively obvious, a sketch in terms of environmental management is needed.

SKETCH OF AN INSTITUTIONAL STRUCTURE
FOR ENVIRONMENTAL MANAGEMENT

The institutional structure for environmental management will logically consist of legislative, executive, and judicial elements, but we shall be concerned mostly with the first two. The judiciary's entrance into the environmental field is already well advanced. In some ways the judiciary is providing a useful transitional instrument for questioning the traditional executive agency handling of environmental issues. Under the leadership of such lawyers as Joseph Sax,[26] groups have been able to call public agencies to account, and individuals whose rights have been violated have been able to seek and obtain relief. There are indications, however (the Supreme Court refusal to give standing to the Sierra Club in the Mineral King case, for instance), that the courts are near the end of their capacity to assist in resolving social choice questions. By its very nature the judiciary must focus on the procedures followed rather than on the substantive aspects of decision making. We can take issue in the courts with the executive agencies' handling of environmental matters, but we must finally face up to making the choices, and that involves both the legislative and executive sides of government.

It will be easier to sketch out the executive side of a metropolitan or regional environmental management structure first. We have already indicated that the technical trade-offs among residual physical forms and the economics of the regional system are such as to demand one integrated

[26] Joseph L. Sax, *Defending the Environment* (New York: Alfred A. Knopf, 1970).

executive agency capable of capturing these effects and economies. No particular problems arise in any such environmental management agency (EMA) beyond the following:

1. Should it be a public corporation chartered by the state, or a governmental agency, either state or locally created?
2. Should it try to manage a whole watershed?
3. Who should control the agency?

The answer to the first question seems to be indeterminate. Either agency could run a monitoring service, make economic and technical calculations of alternative ways of meeting given ambient standards, draw up schedules of effluent charges, initiate punitive action against violators, and advise the policy makers of potential gains and losses and their distributional effects. How to set up the EMA seems to depend on the peculiarities of the local or state situation and not on any conceptual differences between the two methods.

The second question may similarly be begged—how big is the "whole" watershed? It is worth noting, however, that an EMA could work either way. If it has an entire watershed, its options concerning the use of the waterway for disposal of residuals may be expanded (to include low-flow augmentation, for example). If the EMA does not control the whole watershed, then it should have to meet exogenously determined water quality standards, either at its regional boundary or throughout its reach of the river. The former is preferable, since the EMA may want to use the river for disposal and meet its downstream standards by instream aeration or other means. Exogenous standards that close out such uses will unnecessarily restrict the range of management options.

The third question raises the social choice issue. It is abundantly clear that the technical officials, the professionals who run the EMA, will take an overall efficiency point of view. They can prepare long-range plans under different sets of assumptions. What they lack is a policy body that can choose among the plans and can ask that other sets of assumptions be used. This is where the legislative body composed of district GPRs can be useful. It will quickly be seen that using the GPR approach might result in a legislative body of several hundred people for a major metropolitan area. This prospect frightens many, though the reasons given for the fear do not appear substantial. The main fear is that the body would be "unmanageable." There are some normative overtones to that fear, readily identified by asking, "Unmanageable for whom?" Legislative bodies of such size manage themselves very well. (Moreover, if the area becomes very large, obviously the GPRs needed are not the district ones, but the state representatives from the affected area.) A more reasonable way to address the issue is to ask who controls the agenda. The answer is clear; the EMA will, in large measure, by the options it generates for consideration. Indeed, the emphasis should be put on how the process of option generation can be opened up in a nonfrivolous sense. We are currently exploring the ways in which an optimizing model of residuals management can be used in a legislative setting. The essence of the idea is to break out of the model the distributional consequences (good and bad) of particular management options, then to allow legislators to build up coalitions as a result of mutual acceptance of constraint sets.[27] Legislative use of such management models may provide one way by which optimal solutions of determinate problems may be evaluated in terms of social choice.

The district representatives will, of course, be interested in the environmental implications of each proposed solution for their districts. They will also note the financial implication of each solution and weigh it against the overall tax situation in their districts. Some representatives will opt for high environmental quality, even at the expense of higher taxes. Others will accept a deteriorating environment, if necessary, to hold tax rates down. Both will be reflecting constituent preferences. A plan will be adopted; people will move. Is it likely that persons holding contrary preferences, assuming they move, will move into, or out of any district? Both the Tiebout model and we suggest they will move out. Hence this population shift will not affect our solution.

What if one GPR has misjudged his constituency? If he has, *and if elections are frequent*, he will be defeated and his successor will want to undo what he has done. What does this do to the continuity of planning? It will clearly upset it unless an additional rule is made: no execution of a plan until it has been passed in two (or three) legislative sessions. Such rules were once usual in legislative bodies and were used for just such purposes, to ensure that preferences were correctly interpreted on controversial issues.[28]

[27] This concept is elaborated by Russell, Spofford, and Haefele in "Environmental Quality Management"; it will be discussed in greater detail in chapter 5.

[28] For example, the British Parliament Act of 1911 (1 & 2 Geo. 5, c13) which provides for passage by Commons in three successive sessions, extending over at least two years, of any bill voted down in the House of Lords.

If elections are not frequent and/or if one vote can put an area on an irreversible course, then there is no alternative for people who disapprove of the decision except to move. Even under these conditions, which are the ones that obtain in most instances, the extent of political protest, court suits, and civil disobedience suggests that many would rather fight than switch even in such a hopeless case.

Frequent elections might, in particular instances, be an inducement for moving *into* an area rather than out.[29] An historic, picturesque, but soon-to-be-destroyed area might be saved by a movement in by those determined to save it, encouraged by the fact that their votes could swing the balance. Political action *and* moving could be teamed in such cases (remembering, however, the assumption that all issues are encompassed by one GPR) so that one election could turn a district around on many issues.

The operation of such an institutional scheme raises some worries, however. Perhaps a "residuals sink" will be created in one area, either because the economics and technology favor it, or because a majority coalition is built up that finds it mutually advantageous to create it, or perhaps both. How are the minority districts in this situation to be protected? Although such an event may not be likely, it could possibly occur. If it does, then by definition the minority could not "trade out" of the situation, since the situation results from there being a majority-sized "core" in a game theory sense—i.e., a majority coalition no member of which can be made better off by getting out of the coalition.

One defense is to accompany the creation of the new structure with certain minimum environmental standards that are inviolable. Residuals sinks would simply be prohibited, probably by a higher-level government. Another defense would be to assess costs in relation to levels of relative quality, so that people at the lowest quality levels pay nothing or are paid while those at the higher quality levels pay progressively more.

A second problem might occur if the combination of frequent elections and the rule that some plans must be passed twice results in impasse or oscillation. Nothing could be passed. While we do not regard this problem

---

[29] We acknowledge that the strength of Olson's "logic of collective action" could operate against such moves. Since the total economic gain would accrue *only* to those who made the move (and not to all persons desiring to save the area), the idea is still reasonable.

as significant in a practical sense,[30] it is possible. Safeguards could be provided against such contingencies by allowing the agency head to go to the next highest level of general government (usually the state) for approval of a plan if no plan can be adopted locally. This is routinely done in certain cases now.

Another problem that will remain is what happens at the boundary line. We have already alluded to the issue in the water area, but it also occurs in air and in solid waste disposal. Having an EMA will not preclude the necessity for exogenously determined ambient standards at the boundaries; it simply greatly reduces the number of boundaries and simplifies the problem of setting the standards. We may go further and say that the concept of an EMA is instructive to the issue of levels of government in environmental management. Since the rationale of an EMA is to take advantage of technical and economic efficiencies inside a given boundary, it follows that no higher level of government should specify means, only ends. In other words, it may make sense for state or federal agencies to specify an ambient standard (minimum national levels or a ban on heavy metals, for example), but it will not make sense for those governments to prescribe *how* the ambient standards are to be met (by, say, secondary treatment everywhere). The point is well worth making, since the prevailing trend of policy at the federal level is increasingly concentrated on enforcing certain methods of treatment on a national basis instead of on emphasizing the result wanted. This procedure is not only grossly inefficient, ignoring the preferences of the people affected, but probably will not work.

---

[30] Note, on the other hand, the present impasse in both transport and power projects under a governmental structure heavily protected from voter preference.

5

Representative Government and Environmental Management

INTRODUCTION

The following drama was played throughout the land in one form or another during the decade of the sixties. An announcement is made in local papers that a dam has been proposed upstream from central city which will (1) control floods, (2) ensure an adequate water supply to city residents, (3) provide recreation on the lake created by the dam, and (4) improve water quality in the river downstream. Mr. Cond Citizen may read the announcement with only casual interest, comforted, perhaps, that the public officials are doing something worthwhile for a change.

As time goes on, he is puzzled, even bewildered or angered, by a mounting controversy over the dam. The dam is attacked by the people who live on what is to be the bottom of the lake created by the dam. It is attacked by some people who sneer at "flatwater" recreation and speak rapturously of the existing "whitewater" where the dam is to be placed. It is attacked by fisherman who talk about spawning grounds, by biologists who decry artificial lakes and predict silting, by economists who talk about floodplain restrictions, and by engineers who mention reaeration of water as a cheaper method of improving water quality.

Reprinted with minor revision by permission of the Westwater Research Centre, University of British Columbia. The paper was originally presented at the Seminar on the Institutional Arrangements Associated with Planning and Implementing Water-Quality Management Decisions, Victoria, British Columbia, 30–31 March 1973.

117

Mr. Citizen, if he does not screen out all this "noise" in his information channels, may well think, "They must have thought of all that." Mr. Bono Bureaucrat, who did not think of all that, is even more bewildered by the clamor than was Citizen. But Bureaucrat is resourceful. Nobody is going to finger *him* with the blame. He is, after all, only following the law, and besides, if other people want to get into the act, elementary strategy suggests that one should open the stage door and let them in. Once they are on the stage, they will be as committed as anyone to making the act succeed.

So with one hand, Bureaucrat hands the hot potato to the politician, while, with the other, he opens the door to additional interest groups that want in on the action.

Mr. Selwyn Politician, who was born with infrared vision, can spot a hot potato before it begins to glow. He swiftly fields the handoff from the bureaucracy, writes some resounding phrases about the "need for public participation," "multiple use," and "balanced programs," and lobs the potato (now radioactive) back through the bureaucratic transom. For good measure, he creates an environmental agency coequal to the agency proposing the dam in the first place, gives no one final authority, and wishes the judges good health.

The resulting problem can, perhaps, be unfolded by the following three questions:

1. How do we achieve technical efficiency in water quality management?
2. How do we choose the goals for water quality management?
3. Whom do we involve in the process of water quality management?

It will be useful to start in the area of greatest knowledge and move from it gradually out into *terra incognito*.

TECHNICAL EFFICIENCY

Although there is still substantial argument over the means of accomplishing some of the objectives, it remains true that we recognize the necessary conditions for technical efficiency to be:

1. a geographical area that will allow for control, i.e., a watershed, a tributary, or a reach on which boundary conditions can be specified and met;

2. a criterion for choosing among mutually exclusive projects, e.g., benefit-cost in some form;
3. the absence of any bar *against* certain technical options and, conversely, the absence of any external fiscal incentives that bias the technical choice *toward* certain technical options.

These conditions may be thought of, respectively, as our geographical handle on the problem, our economic handle, and our functional handle. To the degree that any of them are defective, in similar measure will we fail to achieve technical efficiency. Unfortunately, as we all know, all three are defective to some extent.

We have just begun to understand how limited the hydrologic definition of the management area is. Not only is it apt to range so widely that it brings in problems (and people) of minimal interest to the management agency, but it also is at odds with the geography of two other interrelated problems—the management of gaseous and solid wastes. These problem-sheds will rarely fit the geography of the watershed, yet there may be efficiency gains to be made by considering all three residuals in one management system. The efficiency gains may be significant in an area like Washington, D.C., and the lower Potomac, whereas tying the whole Potomac basin into such a management scheme may make little sense at present, the demands on the river above Washington being minimal.

Our economic criterion, usually some form of benefit-cost analysis, suffers from well-known vicissitudes on efficiency grounds (for example, choice of discount rates) but is attacked with even heavier artillery when we get into distributional questions. The criterion remains, however, a potent tool by which to make aggregate, economic evaluations of mutually exclusive alternatives. That is a large enough benefit to confer on us, and we should not, as the large federal bureaucracies in Washington are trying to do, put more burden on the tool than it can accept. It will not tell anyone which distribution of costs and benefits is "best" nor does it provide a framework to evaluate noneconomic costs or benefits.

It is notorious that management agencies are seldom allowed to look at technical options in an unbiased way. Governmental programs typically give grants-in-aid for specific technologies, be they sewage treatment plants, dams, or whatever. Rarely, perhaps never, has an agency been able to review all technical options with a clear eye. The fiscal calculation of matching grants blurs every vision, even the strongest. Leading the solution toward a particular technology would, of course, be forgivable were there any rationality or larger optimization being served. None comes to mind.

Yet the lack of perfection in our ability to choose efficiently should not be taken too seriously. It remains true, as Kneese, Davis, and others have demonstrated,[1] that we know how to manage river basins so as to make everybody better off. In practice, we are still below, probably well below, the kind of efficient management we know how to achieve. There are several reasons for this failure. Water quality management is imbedded in a legal and cultural heritage that values technical efficiency less than some other things. Add to that the fact that technical efficiency would displace some people and some areas from a power and influence position they value very highly. Finally, consider the possibility that the legal and cultural context within which these matters rest is, itself, in a pathological state at the moment. The broader institutional structure of governance, particularly in the United States, is inadequate to the tasks that water quality management or, more generally, environmental management, has placed upon it. How that came about leads us into an examination of the goal-choosing mechanisms available to us.

CHOOSING GOALS FOR MANAGEMENT

The common heritage of Canada and the United States consists of what is termed by the present generation of English historians as "the myth of the Anglo-Saxon Constitution." What was a myth in England became a reality on this side of the Atlantic, as both countries established legislative supremacy (in the United States, explicitly limited by a written constitution with a bill of rights patterned closely after the English instrument adopted in 1689) and governments that consciously strove to determine social choices from individual voter preferences. The effort was, in both cases, built from the ground up, from self-governing colonies and territories. Although there are differences, there are greater similarities. One is the great

---

[1] See, for example, Allen V. Kneese and Blair T. Bower, *Managing Water Quality: Economics, Technology, and Institutions* (Baltimore: The Johns Hopkins Press for Resources for the Future, 1968), and Robert K. Davis, *The Range of Choice in Water Management: A Study of Dissolved Oxygen in the Potomac Estuary* (Baltimore: The Johns Hopkins Press for Resources for the Future, 1968).

amount of power lying dormant in legislative halls. In the United States, the power has, in many cases, been curbed by state constitutional amendment (following the legislative scandals of the early 1800s), but the potential to restore it is still there.

During the nineteenth and twentieth centuries, however, the remaining power of the state legislatures has gradually seeped away to other parts of the state government under a well-motivated attempt to get efficient management and has been dispersed to other levels of government. The fiscal power went to the national government as a result of the passage of the Sixteenth Amendment, which authorized the personal income tax. The management power went to special districts, single-purpose authorities, appointed boards and commissions, and to bureaucrats charged with administering laws often vague and unfocused.

In Canada, it seems that less power has been absorbed nationally; certainly in fiscal terms this is so. It may be only slightly overdrawn to say that the Canadian province is a state, but that the American state has become a province.

The reason for this detour into governmental history is this: to the extent that state legislative power has been absorbed or diffused, so has the ability of government to choose, to set priorities, been weakened. The pathology stems from that fact. As one agency gets control of transportation and another control of education, as an appointed board issues sewer permits, another reviews zoning plans, and a third promotes economic development, the whole question of choosing social priorities becomes irrelevant, and we move ahead by chance. Establishing comprehensive planning arrangements and coordination committees, while of interest to a growing bureaucracy, simply compounds the problem of choosing by burying the process from traditional public pressures. (Indeed, the rise of court suits, civil disobedience, and advocacy planning, to be discussed later, results directly from the drying up of the traditional streams of advocacy—the party and the parliament.) If the game is really played in the offices of the bureaucrats or in the hearing rooms of the commission, why bother with parties and legislatures? And, since the bureaucrat and the commissioner are insulated, deliberately, from party pressures, perhaps the court suit is the only way to get into the game at all. So many in the United States have reasoned. If the special-interest lobby can sit in the bureaucrat's office and help write the regulations, then perhaps the only recourse for the opposing special interest is to get into the room also. This method of influencing choices is now so common in the United States that sections requiring "citizen" participation in agency decision making at "every stage" have been written into several of the new federal environmental acts.

Indeed, we may have gone through a whole revolution of decision making without the slightest attention having been paid to it outside the quarters of the interested parties. It has attractions. The new method allows the politician to get off the hook by passing the decision making over to an appointed board or commission. He then can exist on platitudes and good feelings toward all. The political heat is drained off into a kind of infinite sink—a bureaucracy where no one is responsible but all have authority. The interest groups, to the extent they are organized (and Olson has shown us which ones will be),[2] can take their chances on the inside, helping to carve the turkey through informal deals and bargains. The bureaucrat cannot lose; he becomes the indispensible middleman. And, of course, the outsiders and losers always have the courts.

The process just described is going on now in the United States. It is not working. It will not work. It cannot work. Why not? It conforms to many textbook descriptions of the informal power structure of democracies. Moreover, it is consonant with both "pluralist" and "elitist" descriptions of group politics.

It does not, however, conform to, nor is it consonant with, the structure that some very hard-headed politicians put together in Philadelphia in 1787. One of the prime motivations of those men was to prevent a kind of "courtier" (we would say "bureaucratic") government by men not responsive to, or responsible to, the public will. They built well, exceedingly well, out of fire-hardened materials. Hence, today, when management agencies begin to get into important areas, such as influencing or controlling land use and making life-and-death decisions for communities and areas, they find themselves beseiged in the courts. Politicians who practice consensus politics find it impossible to please everyone, and if the legislature did not decide the policy, the one displeased by the policy can often bring the whole administrative process to a standstill in the courts.

Perhaps I belabor the obvious, but the point, no matter how obvious, must be sharpened. Decisions made by general-purpose legislatures, whose

[2] Mancur Olson, *The Logic of Collective Action: Public Goods and the Theory of Groups* (Cambridge: Harvard University Press, 1965).

members are elected, can be overturned in the courts only on the most fundamental grounds of constitutional probity. Decisions made by any other governmental bodies, no matter how carefully designed, can be upset on innumerable procedural and technical grounds by anyone with standing in a court of law. Granting of standing to a broader and broader spectrum of interests (in class action suits, for example) is going forward rapidly in the U.S. court system.

We in the United States are being estopped from doing more and more things. Highways are not being built, dams are not being constructed, buildings are not being put up, industries are not being located, electrical generation capacity is not being expanded. The blockages are not happening everywhere, of course, but they are happening in a growing number of places.

Two reactions are possible in such a period of impasse. One is to recognize why the original system does not allow decision making through bureaucracies and to solve the impasse by returning to the original system. The second, and the one we are unfortunately following, is to attempt to tear down that part of the original system blocking action by bureaucracy and to blast through solutions to the immediate problems. Attempts in the United States to draw these issues up to the federal presidency and to use presidential power (still after 200 years only vaguely defined) will, if successful, fundamentally alter American government in a way precisely analogous to the efforts of the Stuart kings. The attempt will meet, indeed is now being met by, precisely the same response. Myth or no, in the final analysis, we will either govern ourselves or be ungovernable.

My purpose here is not to dramatize these larger issues so much as to use them as the framework for more specific examination of mechanisms for choosing goals. Within the framework one can see the limited role that many sophisticated tools of decision making—systems analytical tools with multiobjective functions, for example—can play. If the fight is over the weights to put on each part of the objective function, whether to allow this or that constraint—when there is not one but many decision makers to contend with, and consensus is impossible—these tools do not suffice.

I add quickly that while they do not suffice, neither are they useless. At RFF, we are now developing management models of just such form, that is, optimizing models with multiobjective function capacity. We are using them, however, in a legislative setting.

Our method has been to construct and experiment with a hypothetical region, modeling its residuals discharge activities and the natural systems that translate those discharges into ambient quality levels. This regional model has been linked to a model that simulates the vote trading activities of a legislature assumed to be responsible for making the decisions about quality in that region. The regional model has been structured to give a great deal of information in physical as well as economic terms to each legislator. We hope to show that information in this form can be used in a systematic way by real legislatures to assist them in arriving at regional policy. Our legislative model is emphatically *not* an attempt to put real legislatures out of business, but simply a device for allowing us to accomplish two things: first, to design the regional model for use in a legislative setting and, second, to compare the policies adopted by legislatures put together along different lines. The basic function of the simulation program is to identify and accomplish the vote trades that would take place in a real legislature.

In an attempt to capture some of the complexity in the environmental issue, we introduce the idea of a preference vector that combines an ordinal ranking and a yes-no vote on a given set of issues. In the legislative model the issues consist of four quality measures of the natural environment and four measures of increased cost resulting from improvements in environmental quality. The former include (1) the level of dissolved oxygen in each reach of the river (calculated in terms of a dissolved oxygen deficit), DOD; (2) increases in the temperature of each reach of the river, ΔT; (3) the level of suspended particulates in the air (measured in micrograms per cubic meter), SP; and (4) the level of sulfur dioxide in the air (measured in micrograms per cubic meter), SO_2.[3] The latter include taxes, unemployment, electricity, and heat.

Arbitrary preference vectors (reflecting differences of tastes, incomes, and the like) were specified for each area. These are based on arbitrary upper limits on each of the eight measures. Thus, one area's upper limits vector is:

---

[3] These measures could probably not be used directly in a political process, although the experience in choosing levels of water quality in the Delaware shows a quick assimilation of technical information by laymen, particularly if the technical measures are related to fish population, recreation potential, and health hazards.

$$
\begin{array}{lll}
\text{DOD} & \begin{bmatrix} 3.0 \\ 5.0 \\ 50.0 \\ 20.0 \\ 1\% \\ 10\% \\ 50\% \\ \$20 \end{bmatrix} & \begin{array}{l} \text{parts per million in reach 4} \\ \degree \text{ F in reach 4} \\ \mu g/m^3 \\ \mu g/m^3 \\ \text{increase} \\ \text{increase} \\ \text{increase to each household} \\ \text{increase per year to each household} \end{array}
\end{array}
$$

DOD, ΔT, SP, SO_2, Taxes, Unemployment, Electricity, Heat

If we allowed area 1 to be our decision maker, these upper limits would be the constraints area 1 would put on the solution of the regional model. Since area 1 is only one of many areas, we wish to construct a social choice process to allow the upper limits vectors of all areas to be expressed. Our preference vectors, one for each area, are designed to do that. For example, area 1's preference vector in response to a current situation is shown in table 17.

Table 17. Area 1: Preference Vector

	Upper limits	Present situation	Preference vector
DOD	3.0 ppm	5.52 ppm	N
ΔT	5.0 °F	9.89 °F	N_1
SP	50.0 $\mu g/m^3$	17.19 $\mu g/m^3$	Y
SO_2	20.0 $\mu g/m^3$	32.69 $\mu g/m^3$	N_2
Taxes	1%	0.0	Y
Unemployment	10%	0.0	Y_3
Electricity	50%	0.0	Y
Heat	$20	0.0	Y

The numerical subscripts in the preference vector indicate the ordinal ranking of three (in this case) measures. Thus we are ranking, by assumption, an upper limit of 5°F heat rise in reach 4 of the river of first importance in area 1, an upper limit on SO_2 of 20 $\mu g/m^3$ as second in importance, and an upper limit of 10 percent on unemployment as third most important.[4] (For ease of computation we do not rank all measures. Elements without subscripts are assumed to be all of equal importance but less important than any subscripted element.)

[4] The ordinal rankings by real actors would clearly change as one or more upper limits were met and/or other upper limits were greatly exceeded. We have not investigated how such ordinal shifts would affect the algorithm, although it is clear that convergence problems might well occur.

Area 1's preferences may be summarized by saying that its citizens are dissatisfied with the present situation in three out of four quality measures, while they are satisfied with the present tax and utility burdens on the area. Since area 1 ranks the environmental measures above the financial ones, however, some additional financial burden would be accepted if necessary to achieve acceptable levels of water and air quality.

All other areas were assigned upper limits vectors and ordinal ranks to three or more measures. Using those vectors, we can display all preference vectors in response to the present situation (table 18). The Y votes on each row are tallied in the far right column. We see a unanimous approval of the financial situation but much disapproval of the present quality of the air and water. The stage is set, assuming our preference vectors are such that not all can be met simultaneously, for some sort of social choice process to be invoked. Since the number of possible "solutions," that is, technically feasible alternatives, may be said to be almost infinite, the social choice process cannot be simply a blind groping for a solution acceptable to some given percentage of areas, for the process would prove to be inefficient and the "solution" ambiguous at best. Neither can we, without throwing away the concept of social choice, simply stand with arbitrary "solutions" that do not reflect the preference vectors in the hope that the "objectivity" of any one of these solutions will cow the residents of our region into acceptance.

In a real-world situation, whatever official has charge of the model (and a computation budget) might be tempted simply to meet the first- and second-ranked upper limits of enough areas to ensure a majority (assuming there existed some political body through which these votes could be registered). Lacking any such political body, he might play around with meeting a few more high-ranked upper limits (chosen on the basis of judicious knowledge of which areas could be most difficult if their limits were not met) and "balance" the additional limits against protests from special-interest groups, industrial and environmental.

It may be useful to explore in some detail what procedures could be used to find a solution based on the preferences of the twenty-five areas. The first method is to see if all preferences (upper limits) of all the areas can be met simultaneously. If so, clearly it is Pareto-optimal to meet them. Since such a happy state is unlikely in the real world, we have set the upper limits vector so that it is not attainable in the regional model either. If all preferences cannot be met, then whose should be met, and in what order?

Table 18. Vote Matrix, Present Situation

Issue	\multicolumn{25}{c}{Area}																									Tally of Y votes
	1	2	3	4	5	6	7	8	9	10	11	12	13	14	15	16	17	18	19	20	21	22	23	24	25	
DOD	N	N	Y	Y	Y	N	N	Y	Y	Y	N	N_5	Y	N	Y_3	N	N	N_1	N	N	N	Y_1	Y_1	N_2	Y	11
ΔT	N_1	N	N	N	Y	N_3	N	N	N	Y	N_3	N_4	N	N	Y	N_3	N	N	N	N	N_2	N_2	N	N_4	N	3
SP	Y	Y	Y	Y	Y	Y_2	Y_2	Y	Y	Y_2	Y_1	Y_2	Y_1	Y	Y_2	Y_2	Y_2	Y	Y	Y_1	Y_3	Y	Y	Y	Y	25
SO_2	N_2	N_1	N	Y	N	N_1	N_1	N	N	N	N_2	N_1	N_3	Y	Y_1	N_1	N_1	N	Y	N	N_1	N_4	N	Y	Y_1	6
Tax	Y	Y	Y_1	Y_2	Y	Y	Y_3	Y_1	Y_2	Y	Y	Y_3	Y_2	Y	Y	Y	Y_3	Y_2	Y_2	Y_2	Y	Y_3	Y_2	Y_3	Y	25
Unemployment	Y_3	Y_2	Y_2	Y_1	Y_1	Y	Y_4	Y	Y_1	Y_1	Y	Y	Y	Y_1	Y_4	Y	Y	Y	Y_1	Y_3	Y	Y	Y_3	Y_1	Y	25
Electricity	Y	Y	Y	Y	Y_2	Y	Y	Y_2	Y_3	Y_3	Y	Y	Y	Y_2	Y	Y	Y	Y_3	Y_3	Y	Y	Y	Y	Y	Y	25
Heat	Y	Y_3	Y_3	Y	Y_3	Y	Y	Y	Y	Y	Y	Y	Y	Y	Y	Y	Y	Y	Y	Y	Y	Y	Y	Y	Y_2	25

A second method is to meet each area's upper limits by using each set separately as constraints on the regional model. This accomplishes two things: it allows us to make sure that each area's upper limits preferences are internally consistent,[5] that is, can be met simultaneously; and it allows us to see the kind of overlap or complementarity between one area's upper limits preferences and those of another area.

A third method is to pay more attention at the outset to the ordinal ranking of the different measures by each area. For example, we might try to meet the first- and second-ranked preferences of all areas, or all first preferences, then all second preferences. Were we thinking about a strong party-oriented legislature or council, "all" might be replaced in the preceding sentence by "majority party." If "all" were possible, the "all" solution might have appeal over the "majority party" solution, but would not necessarily be chosen by the majority party if meeting the minority's first- and second-ranked preferences meant giving up on the majority's third- and fourth-ranked preferences. How the majority party acted would depend very much on what powers of retribution the minority party had, when the next election was coming up, and/or other political factors exogenous to our consideration here.

Any of the preceding methods could be employed in connection with the regional model presented, but they do not get us very far in terms of the reality of conflicting preferences and the loose party structure characteristic of most U.S. legislative councils. Although replicating all the complexities of U.S. legislative procedure and structure would be impossible, we can adopt a method that will replicate an important element in them, namely vote trading at the coalition-formation stage.

The essence of vote trading is giving up on one issue to gain another issue valued more. The basic idea of the vote-trading algorithm is to add constraints to the present situation so that N votes are converted to Y votes in some efficient, nonbiased way. Vote trading is efficient for this purpose because it focuses attention on high-ranked N votes (these are the upper limits violations of most concern to the area). They are the upper limits the areas want most to be put in as constraints on the regional model. With vote trading, however, such a constraint can be put in only if the area that

[5] An area's preferences do not have to be consistent, since it could have "if not this, then that" preferences in it. Identifying these beforehand will be useful, however, in a real-world situation.

wishes to put it in will accept (also allow to be added as a constraint) another upper limit on another issue desired by another area. Constraints are put into the solution, therefore, in pairs.[6]

Since vote trading was explained as giving something up for something of higher value, what is it that each area gives up by this trade? To illustrate, let us pick out a vote trade from table 18 and see what happens.

	Area 13			Area 24		
	Upper limits →	*Present situa-tion* →	*Prefer-ence vector*	*Prefer-ence vector* ←	*Present situa-tion* ←	*Upper limits*
DOD	3.5	3.4	Y	N_2	3.4	3.0
SO₂	80	103	N_3	Y	38	40

The result of the vote trade shown is to put in, as constraints, area 24's upper limit of DOD ≤ 3 on reach 2 of the river and area 13's upper limit of $SO_2 \leq 80$ in area 13. What area 13 gives up is a 3.4 DOD level, an outcome it was happy with since it was below its own limit of 3.5. Area 24 gives up an SO_2 level of 38 in its area, an outcome it was happy with. The results of these trades and further details on the model have been presented elsewhere.[7]

There is no reason to believe that any existing legislature or council would use the trading routine just outlined. There is no necessity for them to do so, since the distributional information may be adapted to a variety of decision paths. For example, in real councils there is often a desire, sometimes even a necessity, to let everybody win. A chairman might ask each member to write down the one constraint he really wants or needs in his district or area. These constraints could be collected and used as the first set of constraints for the model. If the regional model can be solved with those constraints, the effects of the solution could again be examined and one additional constraint added by each member. This process could be con-

---

[6] There is nothing magical about paired constraints. Three or more areas could agree on a constraint before it was put in. Paired constraints are simply easier, computationally, to use in the vote-trading algorithm.

[7] See Clifford S. Russell, Walter O. Spofford, Jr., and E. T. Haefele, "Environmental Quality Management in Metropolitan Areas" (Paper presented at the International Economics Association Meeting, Copenhagen, 19–24 June 1972). Publication forthcoming.

tinued until the regional model fails to solve. The last solution could be adopted or used as a starting solution to trades or bargains.

Bargaining could occur either after the process described in the preceding paragraph was completed or *ab initio*. In general terms, bargaining can take place whenever two or more council members perceive that some constraints are in direct conflict. In such instances, real council members may wish to bargain, that is, to agree to slack off their constraints slightly if the other side will slack off its constraints also. In doing so, they may wish to use the regional model to help them find the most agreeable bargain—one with minimum change in both constraints, one with minimum variation in percent rise between the two, or whatever other criterion is agreed upon by the pair of bargainers.

It will quickly be seen that such constraint slackening will probably come late in the game, since no one will be willing to change his upper limits until it is clear either that they cannot be met as they stand or that a majority favors a solution in which someone's upper limits are greatly exceeded.

The actual use of the regional model, therefore, may vary considerably from one group to another. Some strictures on use are, however, apparent. For example, the distributional information has potentially explosive political content. That fact can be used to advantage for good or ill. Suppressing the information may be politically expedient to some but can rarely, if ever, be either ethically or legally justified when taxpayers' monies are being used. It may not be desirable, however, to have all experimental runs made public so long as they are shared by all the council members. (The potential for mischief by a clique within the council or legislature, a committee for instance, would be very large. Such a group could construct a very biased solution.)

Another stricture on use relates to the course of action to be followed once unanimity has broken down, as it inevitably will early on in any real situation. One can imagine that in a strong party legislature or council the majority party might well wish to give priority to meeting the constraints of its members, with minority preferences to be met later if at all. This method is not as bad as it might appear, so long as the distributional impact of majority deliberations is known to all council members, *and* the minority party has equal access to the regional model to design counter-solutions. These latter solutions will, of course, attempt to meet minority preferences *plus* improving the lot of a sufficient number of the majority members to place the majority solution in jeopardy. That process will, in practice, turn out not too differently from our trading solution.

In more nonpartisan situations, the idea of coalition building may appeal as an equalitarian procedure that may be able to reach a dominant solution in the absence of cycling.

In any event, there will be some pressure on the operators of the model, who must be like Caesar's wife. Sometimes both majority and minority party staff will be needed to ensure ease of access, shared results, and general trust. There will be opportunities for technicians to facilitate a solution or to obstruct one by purely technical means that could go unnoticed by the members or their staffs. These opportunities and dangers are present in many public service posts and cannot be completely eliminated. Just as both sides in disputes have for centuries employed lawyers, so they now must employ programmers, economists, and systems analysts if the technician's temptation to play God (always for the public good, of course) is to be minimized.

In the context of our discussion on choosing goals, a regional residuals management model is but a halfway house on the route to solving our problem. While it does treat residuals in an interrelated fashion and does construct a social choice from representative preference, it does not choose boundaries or relate environmental issues and costs to other issues and their costs.

The latter issue is not too significant in the real world, *so long as the same people who sit in this legislature also decide the other issues.* A representative, in other words, can rank issues only if he has access to all the relevant games being played. If a man serves only on the school board, he does not have to judge between schools and environmental improvement. If he sits on both boards, he may have to choose between them, given the inevitable budget constraints faced by all governments.

Thus, since it is unlikely that we will soon put together the dismembered general-purpose legislature (i.e., abolish all special boards and commissions), it may be necessary to change our habits of selecting different people to serve on each of them. Such a course has salutory effects both on citizen participation and on the boundary problem and is best discussed under our third and final question.

WHO PARTICIPATES?

I earlier outlined the growth, indeed the necessity of growth, of citizen participation in the bureaucratic governance process. Environmental groups, who feel they had previously been outside the decision process, have come inside with a vengeance. I have also suggested that this new emphasis on citizen participation, on widening the numbers of interests and groups, is doomed to failure, and have indicated why. The bureaucrat's room simply is not big enough to accommodate all, and he has no criterion (nor can he have) by which to judge among competing claims for his attention.

Let me support this assertion by calling attention to the present plight of the U.S. Corps of Engineers, as they attempt to broaden their evaluation process to include the economic, social, and environmental effects of their projects.[8] These guidelines respond to Section 122 of Public Law 91–611 (River and Harbor and Flood Control Act of 1970), which asks, among other things, that the Corps evaluate "disruption of desirable community and regional growth." The Corps' attempts to comply would have their comic aspects were the issue not so serious. One example: In a small project, within one county, the Corps has several contending, legitimate representatives of portions of the citizenry, saying different things. How is a bureaucrat to judge, assess, evaluate, or weigh these statements? What incentive, other than human compassion and professional pride, does he have for even making the effort? What recourse do those who lose because of his judgment have?

The immediate impact of this kind of dilemma has been to bring project evaluation to a halt while the Corps struggles in the flypaper of such evaluation. The long-run implication, however, is that questions of *quo warranto* and improper legislative delegation of authority will be raised. The courts have always held that legislative authority, being itself delegated from the people, cannot be delegated.[9] Although we have not yet reached that stage, we are verging upon it when we give appointed officials powers to control land use, economic growth, and environmental quality by their actions in water management. Our present tendency, buttressed by

[8] See "Guidelines for Assessment of Economic, Social and Environmental Effects of Civil Works Projects," Office of the Chief of Engineers, 28 September 1972.

[9] Locke's famous dictum, "Delegata potestas non potest delegari," is but a restatement of a much older tradition in parliamentary rules.

whatever level of "citizen participation" an agency can muster, cannot stand against the ancient precepts of representative government. (It is amusing to reflect that we might go full circle: political activists, in their search for ways for greater participation, may actually rediscover representative government.)

There is, I believe, no need to vex the Corps of Engineers and other agencies with tasks that they are, with good reason, constitutionally barred from doing. There is an alternative, which has the advantage of dealing with the boundary problem as well. Instead of a citizen having many different men represent him—one on the school board, another in the sanitary district, a third in the city or suburban council, a forth on the planning commission, and so on—the citizen should have one man, a general-purpose representative (GPR), to represent him in all those governmental bodies. The basic reason for focusing all citizen representation in one man (at the local level) instead of many is to make the question of priorities, of goals and resolution of conflicting goals, manageable.

The use of GPRs could be of value in overcoming the environmental boundary problem, by separating the question of the geographical reach of the management agency from the geographical area represented on the policy board(s) to which the management agency is responsible. The details of this idea were elaborated in chapter 4, but some discussion is necessary here. Think of a water control agency or environmental management agency as a public corporation created at the state level with power to operate anywhere (much as the Maryland State Environmental Services agency is set up) *subject to* policy direction and funding by councils composed of GPRs in all districts affected by the problem. Thus one council involves only the GPRs in the airshed, another council (overlapping, no doubt) is composed of those GPRs in the watershed.

The implementation involves nothing more complicated than taking two votes in the hall, one of brown-eyed people, the other of brown-eyed and blue-eyed people. It is a building-block process. As the airshed expands, as it will if the city grows, districts could be added automatically on technical evidence that new areas are being affected.

It will be obvious to many familiar with local government that what I am suggesting is in fact already happening in one form or another. Councils of government in the United States are generally composed of elected officials from local jurisdictions. Such men—mayors, councilmen, county commissioners, and the like—already serve as the voting members of transit boards and planning commissions. The GPR is essentially a rationalization and extension of that trend in an attempt to provide maximum chance for the elected official to control what happens in his area and to receive the reward or blame for its happening.

The suggestion would have marked benefits for most professional staffs, for it would give them real representatives to deal with. Few things are more frustrating than trying to gauge community reaction through a board composed of appointees whose knowledge of, and interest in, community reaction is both limited and one-sided. If the GPR system has any merit, it is that it provides incentive for the elected official to know what his constituents want and do not want. It also provides the constituents with an appropriate focus to register their feelings. In an era when the words *participatory* and *democracy* can be joined without any notion of redundancy arising, the incentive and the focus may be worthwhile.

6
Governance of Common-Property Resources

COLLECTIVE DECISIONS AND PUBLIC POLICY

We have been making collective decisions since man emerged as a social animal. Civilized man has developed three ways of making them consciously. In the first way, we decide upon a rule and enforce its strictures on everyone. Such rules are apt to be concerned with man's relationship to other men and to spring from cultural and religious origins. The judicial system evolved from such rules.

A second way we make collective decisions is to elect a man and agree to follow him (so long as he does not violate our rules). The selection of war chiefs and hunting leaders exemplified this method in primitive times. Executive government evolved from these roots.

A third way we make collective decisions is to come together in council, assembly, or parliament. In this way we chanced everything. Here we changed our rules, overthrew our leaders, and consciously grappled with our futures ourselves. Legislative government grew from these convocations.

It may be observed that the law is uppermost in common-property resources when we do not have a goal; that executive government is uppermost when we have an implicit consensus on the goal and are focused primarily on choosing the efficient policies to bring it about; and that the legislature is uppermost when we are trying, consciously, to set goals.

This generalization helps to explain the postwar movement of public policy concerning common-property resources. In the 1950s and early

135

1960s we had an implicit consensus on our goal—economic development—and our attention was focused on developing efficient means. The perfection of benefit-cost techniques and other determinate problem-solving algorithms and their adoption by executive government agencies took place during these years. The general feeling, among professional analysts, was that if the most efficient means of managing a river basin were not chosen it was either that the range of technical options explored was not large enough (single-technology bureaucracies) or that the technical process was "corrupted" by political considerations.

In the 1960s the process of making social choices through technical choice mechanisms broke down almost completely, just as the mechanisms were at their peak of technical competence. The process broke down because the implicit consensus that supported it—agreement on economic development as the primary goal—came apart. People began to focus on the quality as well as the quantity of economic development. Overuse of common-property resources, for example, the air mantle, made many aware that no longer could these resources be ignored in the quest for economic progress.

Once the goal was called into question, the useful role of executive government diminished. The action is now elsewhere. The action is in the courts, because we do not have a goal. In the absence of a goal, we must fall back on rules of general behavior. Besides the general and long unused strictures that provided handles for environmentalists who wanted to slow "progress," lawyers shaped the always malleable common-law heritage into a formidable club with which to challenge executive government agencies. Every local, state, and federal agency finds itself hip-deep in lawsuits and other legal proceedings that effectively hinder action of any sort. It is the genius of the law that when a consensus has broken down, people can use it to find ways of stopping the body politic from continuing to act on the old consensus.

But, while the law can stop us, it cannot start us in a new direction. For that we need a new goal, and a new goal is set through legislative processes that have not yet taken place. I am, of course, aware that exceptions exist to that assertion. The 1954 court decision about racial segregation started us in a new direction, and presidential leadership sometimes determines new social goals. I do not deny the truth of the exceptions, but I do maintain that they are exceptions.

At present, our legislative processes are in a state of disarray, a disarray of at least two dimensions. First, the national legislature has, for reasons that need not concern us here, gradually formed itself into a mirror of executive agencies; that is to say, it has committee counterparts to the executive agencies. As a result, the legislature has lost its ability to act as a committee of the whole, to choose among conflicting priorities, to make judgments across substantive fields, and to operate as the primary marketplace for bargaining, vote trading, and accommodation across a broad range of issues. As a result, the grand trading arena envisioned by Madison has become a series of small trading guilds with all the exclusiveness and insularity that a guild implies.

Second, below the federal level, our problems and people have escaped the boundaries of general-purpose governments. Neither the scope, resources, nor interest remains in many present general-purpose governments to deal with problems of common-property resources. Most people live in one or another of our major metropolitan areas. They move frequently and can have no lasting impact on the small jurisdictions in which they occasionally vote or on the larger metropolitan area that has only informal government. Of course, the actions of every household in its economic role influence the shape and structure of the city. We are, however, discussing collective decisions, not the ordinary reactions to market forces. Higher-level executive government and the law have provided, for some time, the only government available in metropolitan areas. Neither one has a warrant for setting goals.

Some of the resurgence of initiative and referendum voting—both primitive methods of setting goals—can be attributed to the absence of general-purpose legislatures on a scale appropriate to our problems. What that scale is should be a matter of high technical and political interest.

A more radical indictment of the legislature, or indeed, of all three ways of making social choices, could be made. It is that all three systems have broken down because of the inordinate power of corporations and of money to influence outcomes. The power, which can be used to hire the best lawyers, to finance political campaigns, and to provide elaborate lobbying efforts, is undeniable. The adjective *inordinate* is applied when the power is used to produce outcomes not supported by some segments of the population. It is also inordinate if, in fact, it preempts the public space, setting the agenda and denying other problems even access to the public decision-making process.

If this indictment is taken seriously, it draws us back to the rule-making stage and forces us to examine not only our goals, but also our process for setting goals. Reforms that would give equal access to the law, stop private funding of political campaigns, create advocacy planning, and build in an adversary system in large bureaucracies would be in order. We have, indeed, raised a demon when we begin to question how our common-property resources should be governed.

It is the question of rethinking the rules that should be addressed. Specifically, there are three issues that I view as most important. The first consists of getting the adjective *inordinate* out of the power exercised by business interests. The second relates to getting more social costs into private calculations. The third involves putting the rules into a analytical framework.

Business as a Special Interest in the Political Process

It is commonly accepted that there is a basic imbalance between producer interests and consumer interests when both are brought to bear on a public issue. The usual illustration is that of a bill that proposes to restrict a business activity, perhaps by restricting or taxing effluent discharges. The business firms have a direct, immediate, and large financial interest in defeating the bill. Most others have, as individuals, a small, future, and nonfinancial interest in passing the bill. The few firms are organized; the large number of individuals are not. The result is (so the reasoning goes) that such bills are rarely passed or, if passed, are rarely enforced, since the public forgets and the businesses are motivated not to forget.

There is no denying the truth in such illustrations, but it is only a half truth. Perhaps a more pernicious example of business influence would involve business support of a publicity campaign that actually convinced a majority of individuals that the proposed bill, because it would raise the price of goods being produced, would be against the economic interests of all consumers. Such a campaign might not only succeed; it might well be true. "But," cry our environmentalist friends, "the improved water quality would be worth something, too, and no one told the public about that. Moreover, there is no one to tell them except us, and we have very limited means to do so, yet our message is as important as theirs. 'The government,' " so they say, "should subsidize our arguments to give them a visibility equal to the arguments of business." Here is the essence of the "inordinate" power of business problem.

If "the government" tries to do so, how does it know which groups to recognize as "authentic" counterforces? Recent attempts by the Corps of Engineers, the Bureau of Public Roads, the Forest Service, and other federal bureaucracies show an almost ludicrous dilemma building on this issue. No appointed official under our system of government has any warrant to reject or certify groups as "authentic" (this is, by the way, something that every politician can and does do daily). Following the procedural rules becomes the sole refuge of the bureaucrat, and one can predict a large, noisy, and litigious future for this approach to solving the interest-group problem.

The larger question, the power of some interests to dominate the legitimate political process—that is, the legislative and electoral process—needs a far more fundamental approach than "expanded public hearings." It is, I believe, useful to look at these interests as the modern analogue to the power of the church in medieval times. Governments in medieval and early modern times were often dominated by clerics and churches in much the same way that modern governments are dominated by businessmen and business interests. In both cases the collective decision processes were perverted by those whose definitions and goals were more cohesive, more single-minded, and more tenaciously held than the definitions and goals held by the common man. Yet governments did become gradually secularized, and collective choice processes broadened and became more neutral in their workings. How this happened is not a settled historical question, but it is clear that the economic power of the church relative to the state declined.

We are again in an era when the economic power of some corporations is larger than that of some states, when this power, like the power of the church in medieval times, is used literally against sovereign governments. It is worth recalling that when corporate power became strong enough to dominate individual states in the Union (in the nineteenth century) two defenses were used. On the one hand, the federal government began to regulate business, it being larger than any business. On the other hand, citizens of states began to restrict, constitutionally, the power of the state, either forbidding it to act at all or allowing it to act only with a very high majority concurring through special referendum. These latter "reforms," while they stopped the financial hemorrhaging caused by the robber barons, also left our states as crippled partners in the federal Union.

It may be that similar draconian measures will be necessary at the national level now that businesses have grown even larger. It is possible, however, to think of more palatable reforms, such as abolishing private funding of campaigns and reducing the size of corporations. If we assume that such reforms are desirable, how are they to be accomplished within the present structure?

SOCIAL COSTS IN PRIVATE CALCULATIONS

It is also commonly accepted that the private economic calculations of individuals and firms do not account for the social costs (and benefits) that result from private actions. The tragedy of the commons—the overuse of what is not priced—is now a well-understood phenomenon. Several scholars have turned their attention in recent years from documenting the market failure to designing schemes to relate private and social costs. In the water and air quality field, they suggest that a tax on effluents (based on the relative damage done to the environment) would produce this relationship overtly in the business calculation. Regarding land, the suggestion of a site tax or use tax is similarly proposed as a way government could present a developer with a one-time ex ante calculation of the cost to the community of a particular piece of land being developed in a certain way. Other schemes, such as development shares that can be sold separately from the land itself, speak to the distributive aspects of development in a social cost context.

In academic circles most of the attention given to such suggestions centers on the substance of the proposal. Will it accomplish its goal? Can the data be generated economically? Are there unintended side effects of the scheme that vitiate its usefulness? All such questions are important and many yet unanswered. My purpose in raising the subject here, however, is a different one. It is to argue that the schemes, should they prove economical to administer and efficacious in their result, are at best a means to accomplish a policy and can neither substitute for the policy itself nor make it any easier to choose a policy. They simply (no mean task) make it possible to carry out any policy chosen, and at best to carry it out efficiently, that is to say, automatically and without a large supervisory or regulatory bureaucracy.

Though it may be apparent that such taxes provide only a means and not an end, that fact is not often noted. The point is: it is possible to use different reference points on "damages." Shall the river be restored to its level of purity in the year 1600, or 1900, or what? Biological referents are no help, for diversity of species and oxygen levels are not static and timeless absolutes. A choice must be made, and, since the river is not private property, a collective choice must be made. It does not help, moreover, to try to use a willingness-to-pay measure as a referent. Even if we abstract from the preference revelation issue, this approach will not suffice for at least two other reasons. One, the existing distribution of income cannot be assumed away in the social choice context. Two, one would need to know all willingness-to-pay schedules on all other public issues as well, since all compete for the same dollar. In short, there really is a necessity for making collective decisions, one that cannot be assumed away in reality as it can be in the restricted world of economic models.

Putting the Choice of Rules into an Analytical Framework

Our seventeenth- and eighteenth-century ancestors were intoxicated by the vision of a perfected constitution, of a set of rules by which men could govern themselves. They built so well that, as one historian has put it, there is no American political theory because there was no need for one after 1787. It is clear that there is need for one now, and equally clear that it must be built on 1787. Governments, like trees, have roots. We cannot establish a viable political theory unless it is rooted in 1787 and the English constitutional quarrels of the seventeenth century that inspired this Whig republic.

In earlier chapters, I have briefly touched on some of the problems that require political theory answers. They include:

1. The boundary problem for general-purpose governments at the state, regional, and local levels. This, it must be noted, is a different question from that of appropriate boundaries for delivery of public services. The latter is a technical issue related primarily to efficiencies of scale; the former is the social choice question related to who decides what questions.

2. Rules for making decisions about common-property resources. Alternative mechanisms, such as public commissions, referenda, judicial proceedings, benefit-cost calculations, and legislative vote trading will give different answers. Can self-government marry efficiency with equity?

3. The process of legislative deliberations. For example, does the present separation between substantive legislation on the one hand and the appropriation process on the other make sense? Does the separation of the appropriation decision process from the revenue decision process (a split that dates only from the Civil War and came about because of the exigencies of that war) prevent the legislature from controlling the purse?

4. Political parties and participatory democracy. Must we go through a generation of splinter groups, single-issue campaigns, and all the other formative stages out of which two new political parties can emerge or can we build directly on the old two-party structure?

In each of these areas, it has been the problems of environmental quality and management of common-property resources that have vexed the present governmental structure and process. The response of government is twofold. On the federal level, we pass very general laws to "set a national goal" and leave all of the policy making to anonymous bureaus (anonymous, that is, to the ordinary citizen; the affected industries know them very well). On the state and local levels, we appoint boards and commissions to "take these important decisions out of the political arena."

Neither of these actions can be squared with the system of government established in 1787. The latter quotation would have been particularly startling to the confident men who sat in Philadelphia through that hot summer. It may well be that some elements of our governmental system should be changed. We violate no eighteenth-century sensibilities in reasoning about changes that would improve or adapt the system. The founders would probably be aghast that we have changed the formal mechanisms so little despite fundamental changes in the country. We seem, however, to be trying to ignore the roots, to bypass constitutional process while keeping the outward form of the old system, and to pretend we have changed nothing. This self-delusion is an unworthy approach to the 200th anniversary of the Republic.

Appendix A

Electing the President

The 1948 and 1968 presidential elections, when third-party strength threatened to deny a majority of the electoral vote to either major party, brought increased pressure for reform in the system of electing the president. Reform is essential to eliminate the danger of unscrupulous use of their power by electors, to assure that a majority candidate with wide national support is not defeated, and to preclude excessive influence of and, hence, the proliferation of splinter parties.

The present move for reform is led by Sen. Birch Bayh (Dem., Indiana). Senator Bayh's proposal would replace the electoral college system with a nationwide vote count, the winner of which (assuming he received at least 40 percent of the vote cast) would be elected. If no one received at least 40 percent of the vote, a second run-off election between the top two candidates would be held. The purpose of the proposal is twofold: first, to remove the anomaly of the electors (an anonymous group that technically elects the president) and to ensure that the candidate receiving the largest number of popular votes always wins. (It is now possible that one candidate can receive a majority of the electoral vote while another candidate receives more popular votes.)

Analysis of the Electoral System

The electoral system consists of four separable features. Change is needed on two of them, but change on the other two would seriously threaten the existence of the two-party system. (While intuitive support for the two-party system is strong among traditionalists, formal support for the system as a necessary element in a rational social choice mechanism is also provided in "A Utility Theory of Representative Government." See appendix C.)

The four features are:

1. Assigning each state a number of electoral votes (equal to its total representation in Congress), with a majority of electoral votes necessary for election.
2. Awarding a state's electoral vote on a winner-take-all basis.
3. The use of electors to cast the votes of the states.
4. Choosing the president in the House of Representatives should no candidate receive a majority of the electoral vote.

Item 3 can be disposed of quickly. An anomaly from the outset, the use of electors did not proceed according to the intent of the provision and should certainly be abolished. The vote of the state should be cast automatically for the candidate who wins the popular vote in the state.

Item 4 is also relatively simple to diagnose. The problem with the present procedure, by which each state delegation in the House must vote as a unit —i.e., cast one vote—is that it nearly replicates the process that brought on the deadlock in the first place. A more decisive procedure would be to allow each member of the House or of Congress as a whole to vote in his individual capacity.

Item 2 has been the subject of much misguided attention, chiefly by proponents of changes other than the Bayh proposal. In an effort to thwart direct election, some senators have suggested awarding the electoral vote in each state on a basis proportional to the vote. Such a move, like the use of proportional representation anywhere else, practically guarantees the growth of additional parties in any state with a heterogeneous population.

The winner-take-all system, which was quickly adopted by almost all the states, acts much like a single-seat district in promoting the two-party system *in the state*. (Each party must compete actively for a majority of the vote, with no incentive for winning anything less than a majority.)

Item 1 represents the keystone of the electoral system. The effect of requiring a candidate to compete for a majority of the electoral vote is to require him to appeal to all sections of the country. He cannot depend on building up a huge plurality in one state or section to overcome his poor showing elsewhere. The system discounts all majorities over 51 percent in any state. The result is to encourage two center parties *on a national basis*.[1]

Direct election by nationwide count with a run-off in the event that no candidate received the required number of popular votes for immediate election provides additional incentive for minority parties. Clearly, requiring a run-off whenever the leading candidate obtained less than an absolute majority would provide maximum motivation for tiny parties: they would need to get only a miniscule vote to deny election to either major candidate in close two-way races. Making the cut-off 40 percent, instead of 50+ percent, helps but in no way eliminates the problem. It does *not* imply that to be powerful a party must be able to get at least 40 percent of the vote on the first ballot. Quite the contrary, it implies merely that if all "lesser" parties *together* get only slightly more than 20 percent of the vote in a close race— as in 1968, 1960, or 1948, three elections in a quarter century of history— they can throw the election into a run-off.

Run-offs between the top two candidates do not prevent minority parties from influencing the results; *vide* France and the actions of defeated candidates in primaries, especially where run-offs are required. Each small party with influence in some parts of the electorate can bargain powerfully just before the run-off. It is by no means inconceivable that a minority party could in fact designate Cabinet appointees in exchange for its support. In any event, they could well determine which of the two front-runners (and parties) would win.

With such potential ultimate power in the event that as little as 20 percent of the vote be diverted from the majority parties, one can see motives for developing state and regional parties to support every kind of special interest: a revised Townsendite party, extremists of Left and Right, a Black party, and so on. With so strong a possibility of payoff at the national level, lesser parties might be able to elect a significant number of representatives, perhaps forcing coalitions as a means of organizing the House (or conceivably the Senate). With so many groups at this period willing to damn the system from one point of view or another, an electoral reform that

---

[1] It is argued that single-seat districts in other elections will keep a national two-party system together. This is a simple misreading of history. Canada has single-seat constituencies, but a multiparty system; so does France. See Maurice Duverger's *Political Parties* (London: Methuen, 1954), especially p. 223 *et passim*.

would facilitate political polarization poses many dangers. In short, a proposal that sets out to minimize the influence of minority parties would, in fact, do the opposite.

It is perfectly true that the present system can occasionally defeat a majority president. Its ability to do so is one of its strengths.

President Johnson suggested a reform in 1965 (S.J. Res. 58) that would have solved the problems of the present system without destroying the two-party system in the process. It involved abolition of the electors, made the state custom of awarding the electoral vote on a winner-take-all basis a part of the federal Constitution, and suggested resolving any electoral deadlock by a vote of the joint Congress, voting as individual members.

Appendix B

Coalitions, Minority Representation, and Vote-Trading Probabilities

Although there is disagreement on the normative attributes of vote trading in legislative bodies (is logrolling good or bad?), there is little doubt that it exists as one of the ways by which the political process reduces conflict and takes account of intensities of minority preferences. The existence of coalitions of minorities was posited by Madison (in *The Federalist*, Paper no. 10) as fact *and* value. His argument runs along the negative side, i.e., that no tyranny of the majority can exist in the Republic because of the lack of *one* majority on all issues. He neglected (for good reasons) the obverse side of the coin—minorities can band together to pass legislation as well as to defeat legislation. Americans have made good use of vote trading both to pass and to defeat legislation ever since.

There has been no systematic attempt to relate the possibility of vote trading to different coalition patterns, however, perhaps because the task is tedious and the theoretical significance (after Madison) was unrecognized until recently.[1] The advent of the computer has reduced the tedium of the

Reprinted with minor revision by permission from *Public Choice*, Spring 1970 (Blacksburg, Va.: Center for Study of Public Choice, Virginia Polytechnic Institute).

[1] For a general statement, see James M. Buchanan and Gordon Tullock, *The Calculus of Consent: Logical Foundations of Constitutional Democracy* (Ann Arbor: University of Michigan Press, 1962). For the analogy between vote trading and an economic market, see James S. Coleman, "The Possibility of a Social Welfare Function," *American Economic Review* 56 (December 1966): 1105–22.

task, and the work of Riker[2] helps to narrow the task considerably. Riker put back into political theory the notion of the minimum winning coalition (maximum individual benefit to each member of the winning coalition), and this reduces the number of cases that have theoretical significance.

In brief, Riker's theorem states that "in social situations similar to n-person, zero-sum games with side payments, participants create coalitions just as large as they believe will ensure winning and no larger."[3] We can assume that rational coalition formation will make all coalitions of the minimum winning variety *for the purpose* of comparing vote trading in different coalition patterns.

Coalition patterns emerge from the bargaining among members of a legislature, committee, or commission on the issues that come before it. The initial coalitions are formed in the bill-drafting stage and determine initial support for each bill. If, for example, we have five legislators and two issues, only three initial patterns are possible under the assumptions of *minimum* winning coalitions and majority rule. These initial patterns are:

(1) *Case 30*[4]

			Voter			
Issue	1	2	3	4	5	Outcome
A	Y	Y	Y	N	N	Pass
B	Y	Y	Y	N	N	Pass

(2) *Case 22*

A	Y	Y	Y	N	N	Pass
B	Y	Y	N	Y	N	Pass

[2] William Riker, *The Theory of Political Coalitions* (New Haven: Yale University Press, 1962), chap. 2.

[3] Ibid., pp. 32–33.

[4] Case designations are formed by counting frequency with which voters appear in the initial coalitions, thus:

	Number of voters in				
	5 coalitions	4 coalitions	3 coalitions	2 coalitions	1 coalition
Pattern 1	—	—	—	3	0
Pattern 2	—	—	—	2	2
Pattern 3	—	—	—	1	4

This notational scheme was suggested by Elizabeth Duenckel and can be used in designating larger matrices (3 × 5, 4 × 5, 5 × 5) by expanding to the left.

(3) *Case 14*

A	Y	Y	Y	N	N	Pass
B	N	N	Y	Y	Y	Pass

Only seven cases are possible when a third issue is added (3 × 5), eleven cases in the 4 × 5 matrix, and eighteen in the 5 × 5 matrix. Column permutations can be ignored. Table B.1 gives the complete list of cases through the 5 × 5 matrices.

Table B.1. Coalition Patterns

Voting matrix	Case designation
2 × 5	30, 22, 14
3 × 5	300, 211, 203, 130, 122, 122A,[a] 041
4 × 5[b]	3000, 2101, 2020, 2012, 1210, 1202, 1121, 1040, 0400, 0311, 1230
5 × 5[b]	30000, 21001, 20110, 20102, 20021, 12010, 11200, 12002, 11111, 11030, 10301, 10220, 03100, 03011, 02201, 02120, 01310, 00500

[a] A 122 case with duplication of columns.
[b] Some cases have variants if duplicate columns are allowed.

All possible minimum winning coalition patterns having been specified, the probabilities of trading within each case can be calculated once some means of specifying preferences is decided upon. Since the object of the exercise is to compare trading probabilities among cases, the only requirement of specifying preferences is that the method be consistent across cases and matrices. Four such specifications are used to accomplish this comparison. Each method generates some number of preference vectors for each voter.

A *preference vector* is a column vector composed of 0's, 1's, and −1's which indicates whether or not winning on one issue is more important to the voter than winning on another. Thus a 3 × 5 voting matrix (a case 041),

			Voter		
Issue	1	2	3	4	5
A	Y	Y	Y	N	N
B	Y	Y	N	Y	N
C	N	N	Y	Y	Y

in which all issues are passing, might have a preference matrix as follows:

$$
\begin{array}{ccccc}
-1 & 0 & 0 & 1 & 0 \\
0 & 0 & 1 & 0 & 0 \\
1 & 0 & -1 & -1 & 0
\end{array}
$$

where: −1 indicates membership in a winning coalition but a willingness to trade off his vote for another issue on which he is losing;

0 indicates (if winning on the issue) an unwillingness to trade it off, or (if losing) an unwillingness to give up any other issue to gain this one;

1 indicates the voter is losing on the issue and is willing to trade another issue for it. One exception to this notation is explained later.

Examining the preference matrix given, we can identify possible trades by first picking out *trading vectors* (preference vectors which have at least one each −1 and 1). Thus, the *trading vectors* are:

Voter 1	Voter 3	Voter 4
−1	0	1
0	1	0
1	−1	−1

and the only trade[5] is between voters 1 and 4 on issues *A* and *C*:

$$
\begin{array}{cc}
-1 & \!\!\!\!\nearrow 1 \\
0 & \times\; 0 \\
1 & \!\!\!\!\searrow -1
\end{array}
$$

Identification of such trades can be generalized so long as consistent sets of preference vectors are given to each voter.

Preference vectors of the type here being used are the result of combining a given vote vector, e.g., $\begin{bmatrix} N \\ Y \\ N \end{bmatrix}$, with an *ordering* or ranking of relative interest in the issues. Such orderings are traditionally given as $\begin{bmatrix} A \\ B \\ C \end{bmatrix}$, meaning that of issues *A*, *B*, and *C* the voter thinks *A* is most important, *B* next, and *C* least important. The *preference vector* which results from $\begin{bmatrix} N \\ Y \\ N \end{bmatrix}$ and $\begin{bmatrix} A \\ B \\ C \end{bmatrix}$ (assume

---

[5] It should be noted that the trade may or may not be stable. What we are concerned with here is not "solutions" to each "game," but only a test of whether or not any vote-trading possibilities exist.

Table B.2. Preference Vector Sets for 3 × 5 Voting Matrices

Set	Order vectors for generating preference vectors					
Strong ordering (6 vectors per voter)	A	A	B	B	C	C
	B	C	A	C	A	B
	C	B	C	A	B	A
Strong indifference (6 vectors per voter)	A	B	C	BC	CA	BA
	BC	CA	BA	A	B	C
Rotation (3 vectors per voter)	A	B	C			
	B	C	A			
	C	A	B			
Random (8 vectors per voter)	Generate all possible preference vectors directly from a given vote vector.[a]					

[a] For example, given $\begin{bmatrix} Y \\ Y \\ N \end{bmatrix}$ and all issues winning, the logical combinations of 0, −1, 1 are:

$$\begin{array}{cccccccc}
-1 & -1 & 0 & 0 & -1 & -1 & 0 & 0 \\
-1 & 0 & -1 & 0 & -1 & 0 & -1 & 0 \\
1 & 1 & 1 & 1 & 0 & 0 & 0 & 0
\end{array}$$

Some of these vectors assume some trading "outside the game"; e.g., $\begin{bmatrix} -1 \\ 0 \\ 0 \end{bmatrix}$ indicates that the voter would trade off his vote on issue A, but not for any other issue in the game.

all issues pass) is $\begin{bmatrix} 1 \\ -1 \\ 0 \end{bmatrix}$. If only the order vector was changed, say to $\begin{bmatrix} C \\ B \\ A \end{bmatrix}$, the preference vector would be $\begin{bmatrix} 0 \\ -1 \\ 1 \end{bmatrix}$.

Table B.2 sets up four possible ways to generate preference vectors for each voter. The likelihood of any *preference vector* occurring can also be specified, but in the results which follow equal likelihood is assumed unless otherwise noted. (In analyzing real situations, empirical data could be used to make more realistic assumptions about occurrence of certain preference vectors, e.g., one vector twice as likely as another.)

RESULTS

Tabulations of trades "inside the game" and calculation of probabilities is straightforward but tedious if done by hand. A computer program[6] was

[6] Developed by Elizabeth Duenckel, whose perserverance and ingenuity is gratefully acknowledged.

devised which efficiently both tabulates trades and calculates probabilities. As with most combinatorial problems, however, even computer storage must sooner or later give out, so complete results are limited to the 2 × 5, 3 × 5, and 4 × 5 matrices with a few explorations into the 5 × 5 realm (where the base for probability calculations in the random set is 32^5 or 33, 554, 432) and beyond.

The probability of trading is defined as:

$$\frac{\text{number of vector matches}}{\text{number of vector sets}}$$

where a vector set is a selection of one vector from each voter, and the total number of vector sets is V^n where V = number of vectors per voter[7] and n = number of voters. A vector match is defined as any vector set containing at least one trade.

It will be helpful to examine the 3 × 5 matrix first, since the cases are simple, yet show substantial variation typical of larger matrices. Table B.3 gives these probabilities and they are plotted in figure B.1.

Comparisons among cases within a given preference set (i.e., reading down the columns of table B.3) are the relevant comparisons to make, since the levels of probability of any case *across* preference sets (i.e., the rows of table B.3) are artifacts of the preference sets. It is clear, however, that regardless of which preference set is chosen, as the coalition pattern moves from strong dominance by one majority to a multiplicity of majorities (*à la* Madison—from case 300 to case 041), the probability of trading increases.

The imputed preference sets can also be used to see what difference adding issues makes to trading probabilities. For example, figure B.2 shows how trading probabilities increase as the number of issues is increased, using the random preference set.[8]

[7] If all voters have an equal number of vectors. If voters' vectors are unequal, then the total number of vector sets is expressed as $\overset{n}{\underset{1}{\pi}} V$.

[8] It should be noted that the number of vectors each person can have doubles each time an issue is added, i.e.,

$$\begin{array}{llll}
\text{2 issues} = & \text{4 vectors per person,} & & \\
\text{3 ''} = & \text{8 '' '' '' ''} & , \\
\text{4 ''} = & \text{16 '' '' '' ''} & , \\
\text{5 ''} = & \text{32 '' '' '' ''} & ,
\end{array}$$

and that the base for calculating probability goes up as the 5th power of the number of vectors (5 voters). The number of times a particular vector match (potential trade) occurs likewise increases.

Table B.3. Probability of Trading
(3 × 5 voting matrix)

Case	Strong ordering set	Rotation set	Strong indifference set	Random set
300	zero	zero	zero	zero
211 (2)[a]	.39	.33	.19	.11
203 (3)	.54	.48	.28	.17
130 (3)	.54	.48	.28	.17
122 (4)	.70	.63	.43	.27
122A (4)	.75	.68	.49	.31
041 (5)	.81	.75	.54	.36

Note: All calculated on equal likelihood basis and rounded to 2 decimal places.
[a] Numbers in parentheses indicate number of traders.

Case 300	Case 211	Case 203
Y Y Y N N	Y Y Y N N	Y Y Y N N
Y Y Y N N	Y Y Y N N	Y Y N Y N
Y Y Y N N	Y Y N Y N	Y Y N N Y

Case 130	Case 122	Case 122A
Y Y Y N N	Y Y Y N N	Y Y Y N N
Y Y N Y N	Y Y N N Y	Y Y Y N N
Y N Y Y N	Y N Y Y N	Y N N Y Y

Case 041
Y Y Y N N
Y Y N Y N
N N Y Y Y

Although the overall level of probabilities is an artifact of the preference set chosen, some additional evidence of variation as issues are added is given in figure B.3, which uses the rotation set. It should be noted that, no matter what preference set is chosen, the probability of trading can only approach unity. There is always one non-trading vector set—when all preference orderings are identical.

CONCLUSIONS

In "real" situations, each legislator will have, of course, only one preference vector instead of several, and the probability of trading in any legislature, commission, or committee will be a function of that one vector set. It may be worthwhile, however, when devising new commissions, special districts, or other decision structures, to take some note of the number of *independent* issues likely to come before such bodies and to examine the

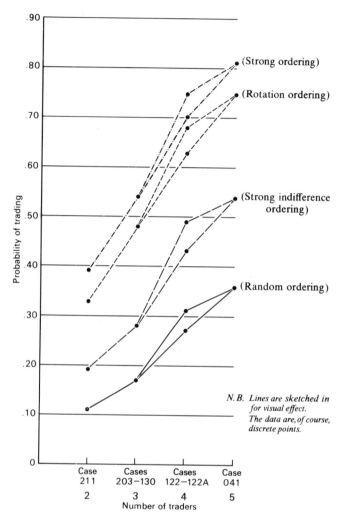

Figure B.1. Vote-trading probabilities (3 × 5 matrix).

probable coalitions and preferences the members of the body are likely to establish relative to those issues. Thus the issue of minority representation can be cast in a new light. If a minority representative is not likely to be needed in any minimum winning coalition, his presence does him no good and is frustrating to him. He is essentially powerless, as he has nothing to trade. Likewise, if the scope of the decision body is restricted to one issue,

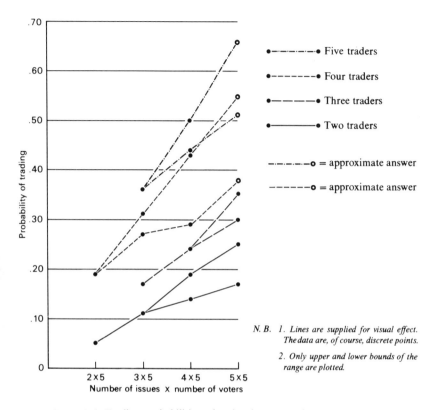

Figure B.2. Trading probabilities related to issues (random preference set).

so that all matters which come before it are likely to be strongly interdependent, then vote trading can play only a small role in decision making. As vote trading is restricted, the probability of *one* dominant majority rises again, with frustrating results for the minority. It also follows, almost without saying, that if the pattern of representation (on the decision body) itself produces *one* dominant majority (i.e., the 300 case), then minority interests are in nowise considered except by the action of altruism, not a reliable defender of minorities.

These considerations may be made clearer by an example. Let us suppose a commission is established to study and make recommendations about water quality in a river. With this its only task, the decisions it takes are essentially mutually exclusive, that is, it faces a set of decisions such that

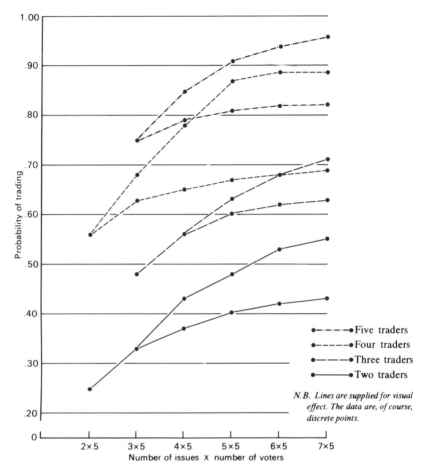

Figure B.3. Trading probabilities related to issues (rotation preference set).

water quality level *A or* water quality level *B or* water quality level *C* may be chosen.

If there are three municipal and industrial water users and two conservation leaders on this commission, the outcome is fairly clear. Concessions to the conservation leaders would take place, if at all, only because of possible effects outside the commission after the recommendation had been made. This concession would have to be made, regardless, and depends not at all on the presence of the conservation interest on the commission. While perhaps self-evident, many boards and commissions function in this fashion, and the equating of "letting minority interests have their say" with democratic process is commonplace.

If the concern with water quality is placed in a somewhat larger context, let us say an interstate agency to manage a river basin, a different pattern emerges. With many issues to resolve, it is less likely that one dominant majority on all issues will occur. For example, we could imagine the following agenda and voting matrix:

Bill	State Rep. 1	State Rep. 2	State Rep. 3	State Rep. 4	State Rep. 5
Water quality	Y	Y	N	N	Y
Effluent charges	Y	Y	N	N	N
Construction of dams	N	Y	Y	Y	N

Under majority rule, the first and third bills would pass, but the plausible assumption of an ordinal matrix:

A	A	B	A	C
B	C	A	C	B
C	B	C	B	A

yields the preference matrix:

0	0	1	1	−1
0	0	0	−1	−1
0	0	−1	−1	1

in which a trade is possible. State Representative no. 5 can trade with either no. 4 or no. 3. All three issues are hence under pressure, and the potential for striking a bargain all can live with is enhanced.

Governments of general jurisdiction have, of course, the widest selection of independent issues on which trades may be struck. The *New York Times* headline of 16 July 1969, "Oil Drilling in Alaska? It Could Determine the Senate's Vote on ABM Issue," evokes the range of potential trading at the national level.

Appendix C

A Utility Theory of
Representative Government

The devolution of economic and political meaning to the individual—hence the identification of the individual *act* as the source of economic and political values—culminated with eighteenth-century rationalism. It was then that the foundation was laid for an economic theory that was to dominate the Western world. The political theory of representative government, codified by the American Constitution, was formulated at the same time. Both theories were devised by men who, as Lawrence Frank puts it, ". . . were persuaded by the 18th century belief in the rationality of man and accepted proposals that emphasized the individual's capacity for acting rationally in pursuing his own self-interest and happiness; calculating his prospective gains and losses."[1]

Since that time, the economic theory based on personal utility calcula-

Reprinted with minor revision by permission from *American Economic Review* 16 (June 1971): 350–367.

Note: Special thanks are due to Allen V. Kneese and Elizabeth Duenckel of the RFF staff for advice, counsel, and much hard work. A note of appreciation is due to Eleanor B. Steinberg, who convinced me not to publish an earlier draft. Orris Herfindahl, Dennis Mueller, and Robert S. Steinberg read the manuscript and made helpful suggestions for improvement. I am also indebted to one of the reviewers for the Farquharson and Shapley references.

[1] Lawrence K. Frank, "The Need for a New Political Theory," *Daedalus* 96 (Summer 1967): 809.

tions has prospered while the political theory based on these same calculations has languished and is now suspect. This is not to say that everywhere
men still trust a laissez-faire market economy while mistrusting representative government. It is to say that economic theory, using the concept of a
competitive market, can explain much about economic systems, both competitive and otherwise, while Anglo-American political theory, having
developed no such counterpart concept, has lost its organizing principle.

The loss came very early, for in one lifetime came the brilliant exposition
of utility-based political theory, the *Federalist Papers*, the not-so-"felicific
calculus" of Jeremy Bentham, and the fatal utilitarianism of J. S. Mill.
Utility-based political theory died aborning, yet the government designed
squarely on its precepts has prospered for nearly two hundred years. Prospered because of its utility foundation? Prospered in spite of its utility
foundation? Do we know?

In 1835, Alexis de Tocqueville was clear it was because of its utility
foundation. "If . . . you do not succeed in connecting the notions of right
with that of personal interest, which is the only immutable point in the
human heart, what means will you have of governing the world except by
fear?"[2] The authors of the *Federalist Papers* were likewise convinced.
Lately the question has shifted considerably and split into two parts: does
representative government really *have* a utility base, and is a utility base
appropriate for the problems of the present? Both questions have important practical as well as theoretical significance and it may prove useful
to review why that is so.

Aggregate measures, whether they be GNP, personal income, net social
return on investment, or whatever, are increasingly disputed because of the
distribution or incidence problem. Who benefits and who pays now occupy
the attention of economists as they turn to questions of public goods. Welfare economics has unearthed no answer to the question of interpersonal
comparisons of utility in the absence of a social welfare function. The generation of such functions has been delegated to the political process, since
it is the formal social choice mechanism. Yet economists have found few
political scientists interested in undertaking the task of explaining precisely
how the generation occurs. Lacking the explanation, economists have tried

[2] Alexis de Tocqueville, *Democracy in America* (New York: Oxford University Press,
1947), pp. 147–148.

their hand at it themselves, notably James Buchanan and Gordon Tullock and Anthony Downs.[3]

Another economist, Kenneth Arrow had already explained how it is *not* done and that explanation has had more impact than the attempts to explain how it *is* done.[4] In brief, Arrow was interested, as were the founding fathers, in determining collective or social choices (choices of public policies or choices among candidates for office) on the basis of individual (voter) preferences. Arrow set up four seemingly reasonable conditions which a social choice mechanism should be expected to meet and found that, as a general proposition, no such mechanism could be devised. No one has disproved Arrow's Possibility Theorem, although some, particularly Duncan Black, have recently raised some hard questions about the relevance of Arrow's conditions.[5]

Arrow had two parts to his Possibility Theorem, and much more ink has been spilt about the General Possibility Theorem than about the Possibility Theorem for Two Alternatives. Arrow proved that the method of majority rule applied to two alternatives does satisfy the four conditions and that the Possibility Theorem for Two Alternatives was, "in a sense, the logical foundation of the Anglo-American two-party system."[6] The restricted theorem has been largely ignored by later writers, perhaps because it was considered trivial from a social welfare function viewpoint.

Arrow did not pursue the implications of his restricted theorem. His concern and the concern of most economists have been for the more general problem and, implicit in such concerns, the analogy with general equilibrium theory in economics. Such concerns and analogy are almost demanded by the logic of individual preference *orderings* and a social preference *ordering* by themselves.

[3] James M. Buchanan and Gordon Tullock, "Public and Private Interaction under Reciprocal Externality," in Julius Margolis, ed., *The Public Economy of Urban Communities* (Baltimore: The Johns Hopkins Press, 1965); Anthony Downs, *An Economic Theory of Democracy* (New York: Harper and Row, 1957).

[4] Kenneth Arrow, *Social Choice and Individual Values* (New York: John Wiley, 1951); 2nd ed. (1963). I am ignoring here the distinction made elegantly by Paul Samuelson between a Bergson social welfare function and Arrow's constitutional function. See P. A. Samuelson, "Arrow's Mathematical Politics," in Sidney Hook, ed., *Human Values and Economic Policy* (New York: New York University Press, 1967), pp. 41–51.

[5] Duncan Black, "On Arrow's Impossibility Theorem," *Journal of Law and Economics* 12 (October 1969): 227–247.

[6] Arrow, *Social Choice and Individual Values*, p. 48.

Such orderings are incomplete postulates in political terms. For example, take the well-known voting paradox ordering

$$A \quad B \quad C$$
$$B \quad C \quad A$$
$$C \quad A \quad B$$

by which, with majority vote as the decision rule, no social ordering can be found. Implicit in that judgment is some voting matrix to which the preference orderings can be related. One such voting matrix is as follows (Y = yes vote, N = no vote, subscripts indicate ordinal ranking of ABC by each voter):

	Voter		
Issue	I	II	III
A	Y_1	N_3	Y_2
B	Y_2	Y_1	N_3
C	N_3	Y_2	Y_1

Here we assume each voter will vote for either his first or second choice, but never for his third choice. This assumption preserves the cyclical outcome for pairwise choices, A preferred to B, B preferred to C, and C preferred to A.

Suppose, however, another voting matrix that does no violence to the ordinal ranking of ABC by each voter:

	Voter		
Issue	I	II	III
A	Y_1	N_3	N_2
B	Y_2	Y_1	N_3
C	Y_3	N_2	Y_1

Now it seems clear that B will win, since voter I is willing to vote for A, B, or C, and in so doing he provides the winning margin for either B or C. He prefers B to C.

Yet another voting matrix with the same ordinal ranking:[7]

	Voter		
Issue	I	II	III
A	N_1	Y_3	Y_2
B	Y_2	N_1	Y_3
C	Y_3	N_2	Y_1

Now it appears that voters I and II will trade votes, defeating both A and B and allowing C to win. This results because both voters I and II have, as a first choice, the *defeat* of an alternative. Thus, though A, B, and C are mutually exclusive alternatives, trading is possible given the appropriate voting matrix.

These cases are commonplace occurrences in the political arena, but they can not be examined by looking at preference orderings alone. In particular, the analogy of a market is particularly inapposite in the last example. There is no market mechanism by which I can directly express the intensity of my dislike for a product. The social choice mechanism of representative government combines a vote matrix with preference orderings to allow expression of negative intensities as well as positive intensities of preference. The use of the combined matrix will be shown later to be of value in understanding the utility mechanism.

Another reason for economists' growing interest in collective or social choices[8] is the increasing problem of externalities in the production process, particularly when these are harmful to the public. Air and water pollution resulting from man's economic activities are the most visible example. If air and water are free goods, or nearly so, then they may be overused because private cost calculations do not include the costs imposed on other

---

[7] Note, however, that now the ranking must be strictly interpreted as intensity of preferences, whether for or against. Thus voter I prefers first, A to lose; second, B to win; and third, C to win.

[8] See particularly Robert Ayres and Allen Kneese, "Production, Consumption, and Externalities," *American Economic Review* 59 (June 1969): 284–297; and Lloyd Shapley and Martin Shubik, "On the Core of an Economic System with Externalities," *American Economic Review* 59 (September 1969): 678–684.

people. Hence, as Buchanan and Tullock have noted,[9] the economic basis for taking collective actions has shifted from opportunities for external economies to conflicts over external diseconomies. The import of this shift is to focus attention on governmental action rather than on corporate action. For, while schemes can be devised (effluent charges, for example) which would internalize these costs, they must be adopted to be effective. Adoption presumes a collective choice and persons hurt by this choice will resist it. If the scheme is rejected, how are we to judge the rejection? Is whatever the political process churns out "right" by definition? Some social scientists have called this trust of the political system into question. Daniel Bell asks for a new political theory to provide a way of choosing between the welfare gains and the welfare losses.[10]

Allen Schick warns that the political system may be defective in a manner analogous to an imperfect market.[11] He notes that since the impact of public goods decisions falls unequally on different groups, the political mechanism, far from providing clear welfare criteria for choice, may produce either too much of a public good (defense?) or too little (environmental quality?). Is there no political analogue to Smith's invisible hand?

In sum, determining whether or not representative government does have a utility base and asking whether or not such a base is appropriate are both revelant areas for study. This paper is mainly concerned with demonstrating the utility base but does offer some thoughts on the latter question.

---

[9] ". . . as people get richer, they need to rely less and less on their neighbors to cooperate in securing the individual benefits of possible joint activities, but they may need to rely more and more on some collective mechanism to prevent themselves, and their neighbors, from imposing mutually undesirable costs on each other. . . . 'congestion' replaces 'cooperation' as the underlying motive force behind collective action" (Buchanan and Tullock, "Public and Private Interaction," p. 69).

[10] "The political tradition from John Locke to Adam Smith paved the way for a new society in which representative government and the free market economy served as the framework for a system of individual decision-making based on self-interest and rational choice. Can one write a new political theory . . . that deals with a service state and a society characterized by a new mixture of individual and communal public and private decision-making units? . . . where joint decisions are to be made, are there clear welfare criteria that justify one choice rather than another?" (Daniel S. Bell, "A Summary of the Chairman," *Daedalus* 96 [Summer 1967]: 699, 977).

[11] Allen Schick, "Systems Politics and Systems Budgeting," *Public Administration Review* 29 (1969): 137–151.

REPRESENTATION AND INDIVIDUAL UTILITY

The link between utility and representation has proved a major stumbling block which hinders our understanding of the utility base of representative government. That representative democracy, as opposed to pure democracy, was necessary for effective government was almost self-evident in the eighteenth century.

James Madison in *The Federalist*, Paper no. 55, expresses the point most succinctly: "Had every Athenian citizen been a Socrates, every Athenian assembly would still have been a mob." Need it be added that Madison's words relate to information costs, revealed preferences, and the lack of a vote-trading mechanism, or that modern proposals that everyone vote on all issues by electronic processes suffer the same defect as the Athenian assembly?

Arguing the necessity of representation on negative grounds does not address the question of the link between individual utility and the representative, however. Furthermore, the contention that the representative will be wiser, more judicious, less swayed by whims of the moment—whatever its truth may be—likewise begs the question. Let us address it directly.

Consider three men as making up a district. Two independent issues are posited as being important for resolution, and the men's positions on these issues are described by the following combined vote and preference matrix:[12]

| | Voter | | |
Issue	I	II	III
A	Y_2	Y_2	N_2
B	N_1	Y_1	Y_1

where again Y is a vote for, N a vote against, and the subscripts are the ordinal rankings of the issues by each man. In this case, all three men rank

---

[12] In this and all subsequent examples, I assume no position on whether a representative should lead or follow his constituency. The individual preference orderings and voting positions can be considered sui generis or as having been formed from the persuasions of prospective or actual representatives.

issue B as more important than issue A, voter I prefers the defeat of issue B to passage of issue A, voter II prefers the passage of B to the passage of A, and voter III prefers the passage of B to the defeat of A.

Two methods exist whereby these men may decide these issues. They may meet as an assembly; in such case, it is obvious that under majority rule, both issues will pass (no trades are possible). Alternatively, they may elect a representative (not one of the three voters). In that case they face a mutually exclusive choice of one of four possible outcomes on the two issues. If we display these as alternative outcomes with the consequences for each voter under each outcome (P = pass, F = fail) we have:

	P	P	F	F
Issue A	P	P	F	F
Issue B	P	F	P	F
Voter I wins	2	12	none	1
Voter II wins	12	2	1	none
Voter III wins	1	none	12	2

Thus, if a representative were elected on a $\left[\begin{smallmatrix}P\\P\end{smallmatrix}\right]$ platform, voter I would win only his second choice (passage of A), voter II would win both his first and second choices, and voter III, his first choice only. Obviously, voter II would vote for a representative espousing a $\left[\begin{smallmatrix}P\\P\end{smallmatrix}\right]$ platform, but voters I and III would be motivated to look for alternatives to $\left[\begin{smallmatrix}P\\P\end{smallmatrix}\right]$. No single alternative is preferred by both voters. We may conclude that $\left[\begin{smallmatrix}P\\P\end{smallmatrix}\right]$ would win over any rival, and that the perceptive aspirant for office will run on this platform.

Sometimes an assembly outcome depends on vote trading in which issues of lesser utility are traded for issues of greater utility. Consider this rearrangement of preferences:

		Voter	
Issue	I	II	III
A	Y_2	Y_1	N_1
B	N_1	Y_2	Y_2

The assembly outcome of this arrangement would be that issue A passes and issue B fails. This results from an attempted trade between voters I and III (which would result in failing both issues), which voter II prevents by giving up issue B (voting N instead of Y), thereby keeping voter I's Y vote on issue A.

Under an election process, obviously the voters cannot directly trade votes. They are again faced with a choice of mutually exclusive alternatives. The alternative outcomes are:

Issue A	P	P	F	F
Issue B	P	F	P	F
Voter I wins	2	12	none	1
Voter II wins	12	1	2	none
Voter III wins	2	none	12	1

Now, if we are to arrive at the same outcome $\begin{bmatrix} P \\ F \end{bmatrix}$ as did the men when meeting as an assembly, the path which the choice process takes becomes crucial. Proceeding as we did before, voters I and III can be attracted to $\begin{bmatrix} F \\ F \end{bmatrix}$ as an alternative to $\begin{bmatrix} P \\ P \end{bmatrix}$. Were $\begin{bmatrix} F \\ F \end{bmatrix}$ to be chosen by one candidate, a second candidate could win on a $\begin{bmatrix} P \\ F \end{bmatrix}$ platform, *assuming no other candidate runs.*

While it is not remarkable that a path can be found, by means of a chain of assumptions, which brings us to the solution reached through assembly trading, it is significant that the chosen path is not unlike a two-party system groping toward positions on issues.

Before formulating rules by which that path can be specified when more than two issues are involved, it may prove helpful to explore outcomes of all ordinal permutations of the $\begin{bmatrix} Y Y N \\ N Y Y \end{bmatrix}$ voting matrix. These are shown in table C.1.

If we view each permutation as a separate district, we can see what happens when voters in different districts are not concerned about the *same* issues.

Suppose that the voters from districts (permutations) 4, 5, and 6 are now considered in terms of three issues, as illustrated below table C.1.

Table C.1. Legislative-Assembly Outcomes
(2 × 3 matrix)

Case 12	Voter			Outcomes	
Issue	I	II	III	Assembly	Repre-sentative
A	Y	Y	N		
B	N	Y	Y		
	Ordinal permutations:				
(1)	1	1	1	Pass	Pass
	2	2	2	Pass	Pass
(2)	1	2	1	Pass	Pass
	2	1	2	Pass	Pass
(3)	1	1	2	Pass	Pass
	2	2	1	Pass	Pass
(4)	1	2	2	Pass	Pass
	2	1	1	Pass	Pass
(5)	2	1	1	Pass	Pass
	1	2	2	Fail	Fail
(6)	2	2	1	Fail	Fail
	1	1	2	Pass	Pass
(7)	2	1	2	Pass	Pass
	1	2	1	Pass	Pass
(8)	2	2	2	Pass	Pass
	1	1	1	Pass	Pass

Note: Case designations are formed by counting the frequency with which voters appear in the initial coalitions. In Case 12, the only nontrivial 2 × 3 case, one voter is in two coalitions, and two voters are in one coalition, hence Case 12.

		Voter		
	Issue	I	II	III
District 4	A	Y_1	Y_2	N_2
	B	N_2	Y_1	Y_1
	C	3	3	3
District 5	A	Y_2	Y_1	N_1
	B	3	3	3
	C	N_1	Y_2	Y_2
District 6	A	3	3	3
	B	Y_2	Y_2	N_1
	C	N_1	Y_1	Y_2

If an issue is not relevant in a district—as, for example, issue C in district 4—we can ascribe to it the third position in the ordinal ranking of each voter even though we ascribe no Y or N position. Since we have not changed the voting pattern or the ordinal rankings, we know the outcome in each district (whether decided by election of a representative or by assembly action) by reference back to table C.1. Thus the representatives from each district, when they meet in a regional assembly, could be shown as

	Representative		
Issue	4	5	6
A	Y_2	Y_1	3
B	Y_1	3	N_2
C	3	N_2	Y_1

If we use majority rule as the decision rule, then the blanks may be replaced, plausibly, by N's, since the outcome is not changed thereby. The resulting matrix

	Representative		
Issue	4	5	6
A	Y_2	Y_1	N_3
B	Y_1	N_3	N_2
C	N_3	N_2	Y_1

has one potential trade (which is blocked by representative 5) and the outcome is to pass the first two issues and fail the third.

Were the nine voters to come together as an assembly, thus

	Voter								
Issue	I	II	III	IV	V	VI	VII	VIII	IX
A	Y_1	Y_2	N_2	Y_2	Y_1	N_1	3	3	3
B	N_2	Y_1	Y_1	3	3	3	Y_2	Y_2	N_1
C	3	3	3	N_1	Y_2	Y_2	N_1	Y_1	Y_2

and the blanks replaced by N's

				Voter					
Issue	I	II	III	IV	V	VI	VII	VIII	IX
A	Y_1	Y_2	N_2	Y_2	Y_1	N_1	N_3	N_3	N_3
B	N_2	Y_1	Y_1	N_3	N_3	N_3	Y_2	Y_2	N_1
C	N_3	N_3	N_3	N_1	Y_2	Y_2	N_1	Y_1	Y_2

again the first two issues pass and the third fails, although the process of trading that accomplishes this outcome is a little more complex. All issues are losing initially, but voter VII must change his vote on issue A to block a trade between voters I and III which would cause issue C to win. Similarly, voter IV must change his vote on issue B to keep issue C from winning because of a possible trade between voter III and either voter V or VI.

While the foregoing illustration may be obvious, it does demonstrate that the process works across districts without uniformity of issues in every district.

Rules for Solution in the 3 × 3 Vote Matrix

Following William Riker in using only minimum winning coalitions,[13] only two 3 × 3 vote patterns need be considered:

Case 030				Case 111		
Y	Y	N		Y	Y	N
Y	N	Y		Y	Y	N
N	Y	Y		Y	N	Y

Each case has 216 ordinal permutations. (Manipulation of Y's and N's is not necessary to exhaust the set of permutations.)

Legislative Vote Trading

Rules for solving the cases considered as assemblies are fairly straightforward. We assume voting stances are known but that relative importance of issues (ordinal ranking) is revealed only by actions. The rules are:

[13] William Riker, *The Theory of Political Coalitions* (New Haven: Yale University Press, 1962).

1. Trades take place if, and only if, they are mutually advantageous.
2. Any trader prefers a higher gain-to-loss ratio to a lower one, e.g., will trade a third choice for a first choice in preferences to a second choice for a first.
3. Any trade can be reversed (canceled) by a third voter if he can offer one of the traders an alternative which is more advantageous to that trader and less damaging to himself than the trade would be.
4. Bluffs and threats, defined as actions which, if taken, would harm the actor, are not allowed.
5. While only ordinal utilities are used, it is convenient to set choice 1 = choices 2 + 3 for all voters.
6. Majority rule is the decision rule.

Solutions determined by these trading rules were assumed "correct" for the purpose of devising rules for selection of a representative.

Rules for Selecting a Representative

Again we assume that voting stances (opinion polls?) are known at the outset. The rules determine the decision path:

1. Start at the nominal outcome, i.e., count the votes. As presented here, that is always at the vector $\begin{bmatrix} P \\ P \\ P \end{bmatrix}$.
2. Select from the remaining seven possible vectors, the outcome(s) which is (are) most advantageous:
   (a) to the two voters "worst off" in the $\begin{bmatrix} P \\ P \\ P \end{bmatrix}$ vector[14]
   (b) or, if the two "high" men are tied, to the one low man and either one of the high men[15]
      (i) if more than one vector choice is possible, choose the one most favorable to the low man[16]
      (ii) if no such vector exists, choose the outcome favored by the two high men[17]
   (c) or, if all three voters are tied, to any pair of voters.[18]

---

[14] See example 1, at the end of the appendix.
[15] See example 2.
[16] See example 3.
[17] See example 4.
[18] See example 5.

3. If no vector can be chosen under rule 2, then the initial vector is chosen by all parties.
4. If a vector is selected under rule 2, then it becomes one party position.
5. One other party position is selected from the remaining six vectors by choosing the vector(s) which adds, and only adds, to the winnings of one member of the first party (in rule 2 above). If more than one outcome is possible, choose the outcome most advantageous to the voter not a member of the first party.[19]
6. The two party positions are matched and the vector selected which would receive the majority vote. (As set up here, this will always be the vector selected in rule 5 above.)
7. While only ordinal utilities are used, it is convenient to set choice 1 = choices 2 + 3 throughout for all voters.
8. Majority vote is the decision rule.

Results of applying these rules to the 3 × 3 matrix are shown in table C.2.[20] Several items in the results need some explanation. First, where a bargaining situation develops (either in the legislative process or in the selection process), the same assumption must be used if the two processes are to produce the same result. This is not surprising, but does mean that it is occasionally possible, particularly in Case 111, for some of the outcomes to be interpreted differently. Second, there are two permutations of Case 030 for which two solutions are possible because of our "choice 1 = choices 2 + 3" rule.

These two cases are the last vestige of the cyclical majority phenomenon. They are testimony to the sturdy truth underlying Arrow's General Theorem.

THE CYCLICAL MAJORITY UNCYCLED[21]

While the paradox of the cyclical majority has been traditionally posed as involving mutually exclusive issues—alternative social states, candidates

[19] See example 6.

[20] All cases were solved by hand, but to avoid the possibility of a shifting premise, computer programs were written and the cases solved again. Fortunately, the two methods give identical answers. Elizabeth Duenckel developed the programs and her help is gratefully acknowledged.

[21] James C. Coleman read an earlier version of this section and made valuable suggestions for improvement.

Table C.2. Legislative-Representative Outcomes
(3 × 3 matrix)

Case 030

Issue	Voter I	Voter II	Voter III
A	Y	Y	N
B	Y	N	Y
C	N	Y	Y

Ordinal permutations (216 possibilities):

	Outcomes Legislative	Representative
100 of which	Pass	Pass
	Pass	Pass
	Pass	Pass
38 of which	Pass	Pass
	Pass	Pass
	Fail	Fail
38 of which	Pass	Pass
	Fail	Fail
	Fail	Fail
38 of which	Fail	Fail
	Pass	Pass
	Pass	Pass
2 of which	Fail or Pass	Pass or Fail
	Fail or Pass	Pass or Fail
	Fail or Pass	Pass or Fail

Case 111

Issue	Voter I	Voter II	Voter III
A	Y	Y	N
B	Y	Y	N
C	N	N	Y

Ordinal permutations (216 possibilities):

	Outcomes Legislative	Representative
132 of which	Pass	Pass
	Pass	Pass
	Pass	Pass
42 of which	Pass	Pass
	Pass	Pass
	Fail	Fail
21 of which	Pass	Pass
	Fail	Fail
	Fail	Pass
21 of which	Fail	Fail
	Pass	Pass
	Pass	Pass

Note: Cases are designated as in table C.1. Cases 030 and 111 are the only two nontrivial 3 × 3 cases.

for one post, considered without side payments—both side payments and independent issues have been introduced into recent discussion. Arrow comments on side payments:

> If, instead of assuming that each individual votes according to his preferences, it is assumed that they bargain freely before voting (vote-selling), the paradox appears in another form. . . . If a majority could do what it wanted, then it would be optimal to win with a bare majority and take everything; but any such bargain can always be broken up by another proposed majority.[22]

We have seen, however, that once vote trading is allowed, the paradox does not hold if issues are independent. That result is not limited to single-peaked preference functions, which Tullock and Paul Simpson have investigated; nor need it come about because of considerations of utility under uncertainty, as in the work of Coleman; nor does it depend on the existence of a pseudo–price mechanism, as posited by Robert Wilson.[23]

To demonstrate this, let us examine the permutation usually considered in a mutually exclusive issue context as an independent issue case (using our rules for legislative trading). Changing the vote matrix to comply with the assumption that each voter votes for his first choice and against the other two, we will have (majority rule):

		Voter		
Issue	I	II	III	*Outcome*
A	Y_1	N_2	N_3	F
B	N_2	N_3	Y_1	F
C	N_3	Y_1	N_2	F

[22] Kenneth Arrow, "The Organization of Economic Activity: Issues Pertinent to the Choice of Market Versus Non-Market Allocation," in *The Analysis and Evaluation of Public Expenditures: The PPB System*, Subcommittee on Economy in Government of the Joint Economic Committee of U.S. Congress (Washington, D.C.: 1969), pp. 47–63.

[23] Gordon Tullock, "The General Irrelevance of the General Impossibility Theorem," *Quarterly Journal of Economics* 81 (May 1967): 256–270; P. B. Simpson, "On Defining Areas of Voter Choice: Professor Tullock on Stable Voting," *Quarterly Journal of Economics* 83 (August 1969): 478–490; James S. Coleman, "The Possibility of a Social Welfare Function," *American Economic Review* 56 (December 1966): 1105–22; Robert Wilson, "An Axiomatic Model of Log-Rolling," *American Economic Review* 59 (June 1969): 331–341.

In order to examine the trading sequence, it will be useful to display the subscripts (rankings) separately and construct a trading matrix:

| | *Voter* | | |
Issue	I	II	III
A	1	−2	−3
B	−2	−3	1
C	−3	1	−2

Each column vector shows a voter's trading desires. Applying our rules for trading, the voter *can* trade any cell with a negative sign but he will, obviously, only trade lower preferences for higher. Thus in this display each preference vector shows a desire and ability to trade either a second or third choice for a first choice. If we choose a first trade arbitrarily, since none is dominant, the trading sequence is illustrated in table C.3.

Since there is no familiar notation to show the trading sequence, it may be helpful to describe the process shown in table C.3. Starting in tableau I, and recalling that columns represent voters and rows are issues, we see that all three issues fail initially. Voters II and III trade votes on issues B and C (shown by the crossed arrows). This trade, if allowed to hold, results in the solution S_1, namely that, voter I would win nothing, voter II would win on his first and second preferences, and voter III would win on his first and third preferences.

This trade is not stable, however, as it is in voter I's best interest (and in his power) to reverse the trade by giving up his vote on issue C to voter II so that voter II does not need to trade with voter III. Hence, the S_2 solution at the bottom of tableau I.

Tableau II is generated by the S_2 solution; the outcomes are now F, F, P, and voter I is shown voting Y on issue C. The negative and positive signs in the trading matrix are changed accordingly and additional trades are sought. There is a possible trade (again shown by crossed arrows) and this time the potential reversal cannot occur. Voter II could reverse this trade only if he is willing to give up voter I's support on issue C. But, since issue C is his first choice, he will not be willing. Hence, S_3 is shown as the result of the trade between voters I and II on issues A and B.

Tableau III is generated from this solution (S_3); all issues are now pass-

Table C.3. Trading Sequence

	Vote matrix				Trading matrix			Solution	Winning preferences		
	Voter I	Voter II	Voter III	Out-come	Voter I	Voter II	Voter III		Voter I	Voter II	Voter III
					Tableau I						
Issue A	Y	N	N	F	1	−2	−3	S_0 Initial	2nd 3rd	2nd 3rd	2nd 3rd
B	N	N	Y	F	−2	−3	1	S_1 (trade)	none	1st 2nd	1st 3rd
C	N	Y	N	F	−3	1	2	S_2 (reverse)	2nd	1st 2nd 3rd	3rd
					Tableau II						
Issue A	Y	N	N	F	1	−3	−3				
B	N	(Y)	N	F	−2	−3 → 1[a]	1	S_3 (trade)	1st	1st	1st
C	(Y)	Y	N	P	3	−1	2				
					Tableau III						
Issue A	Y	N	(Y)	P	−1	2	3				
B	(Y)	N	Y	P	2	3	−1	S_F Final	1st	1st	1st
C	(Y)[b]	Y	N	P	3	−1	2				

Note: ⤢ indicates trade. → indicates a reversal of trade. ↛ indicates a blocked reversal.

Each successive tableau of voting matrix and trading matrix is set up on the basis of trading done in preceding tableau; thus tableau II starts at solution S_2 and tableau III at S_3.

[a] Reversal blocked because voter I can retaliate by withdrawing his support on issue C (last row).

[b] This vote remains a Y because it enables voter I to hold issue B (second row) so that he can trade it to voter III.

ing, and the new trading matrix shows no further trades are possible. Three votes are shown "circled," indicating that they are cast Y as the result of trades. The result does not depend on voter I being willing to trade off *both* issues B and C to get A. In other words, $A \succ B + C$. He trades off C to keep B. When he has B, he can use it to get A, which he prefers to B.

When the solution $\begin{bmatrix} P \\ P \\ P \end{bmatrix}$ has been reached, it is stable unless the rule "choice 1 = choice 2 + 3" is relaxed. If at least two voters think choices $2 + 3 > 1$, the solution could be forced to $\begin{bmatrix} F \\ F \\ F \end{bmatrix}$.

When this permutation appears in a mutually exclusive context (representative rules) the same two solutions appear. Displaying the outcomes as before:

P	P	F	F	P	P	F	F
P	F	P	F	P	F	P	F
P	P	P	P	F	F	F	F
1	12	—	2	13	123	3	23
1	13	12	123	—	3	2	23
1	—	13	3	12	2	123	23

and noting that $\begin{bmatrix} F \\ F \\ F \end{bmatrix}$ is the initial vector, we can choose no other vector under rule 2. Hence (under rule 3) $\begin{bmatrix} F \\ F \\ F \end{bmatrix}$ is chosen by both parties. Only if choice 1 > choices 2 + 3 could $\begin{bmatrix} P \\ P \\ P \end{bmatrix}$ dominate $\begin{bmatrix} F \\ F \\ F \end{bmatrix}$.

Thus the legislative rules choose $\begin{bmatrix} P \\ P \\ P \end{bmatrix}$ and the representative rules choose $\begin{bmatrix} F \\ F \\ F \end{bmatrix}$. This divergence is uniquely determined by the convenience rule, choice 1 = choices 2 + 3. If this rule is relaxed, the two processes choose the same outcome. If choice 1 > choices 2 + 3, $\begin{bmatrix} P \\ P \\ P \end{bmatrix}$ is chosen. If choice $1 <$ choices $2 + 3$, $\begin{bmatrix} F \\ F \\ F \end{bmatrix}$ is chosen. This convergence requires an addition to the representative rules to deal with the following special case when it occurs:

$$\begin{bmatrix} P \\ P \\ P \\ \hline 1 \\ 1 \\ 1 \end{bmatrix} \begin{bmatrix} F \\ F \\ F \\ \hline 2\ 3 \\ 2\ 3 \\ 2\ 3 \end{bmatrix}$$

A more general solution to this problem would require additional specified elements in each ordinal ranking. That was not attempted.

The other nominally cyclical examples (there are a total of 12 in each 216 permutations) come to a single stable solution, both as independent issues (legislative rules) and as mutually exclusive choices (representative rules), and are considerably less complicated than the example presented here.

A caveat from the real world should be issued at this point. The practice of statecraft cannot always depend on legislative vote trading or a two party choice process. When decisions must be made rather separately, so that vote-trading possibilities are reduced, the decision is rarely made by simple majority vote. The role of the nominating committee as a screening device is familiar to all.[24] It was not by accident that nominating committees, party caucuses, the King's Council, and other devices grew up in Anglo-Saxon government to prevent indecision (lack of convergence) in governmental processes. This growth took place, not in the grip of utility theory as did the "democratic" elements of the system, but as a historical reaction developed during the long and bloody struggle for control of the English Crown. The early history of the English Commons shows it assenting to or rejecting a proposed levy. As the initiative gradually passed from king and council to Commons, the utility part of the decision process was confined to adjustments on local issues. Individual members' preferences do not shape the bills involving national (rather than local) interests. The latter bills were (and are) shaped by party leadership.

Walter Bagehot, writing near the middle of the last century, is still instructive on the point:

> . . . the principle of Parliament is obedience to leaders. . . . The penalty of not doing so is the penalty of impotence. It is not that you will not be able to do any good, but *that you will not be able to do anything at all.* If everybody does what he thinks right, there will be 657 amendments to every motion, and none of them will be carried or the motion either.[25]

Moreover, there are times when even party leadership will not suffice. Writers from Aristotle—"the nature of a *polis* is to be a plurality"—to

[24] Lewis Carroll's struggles with majority rule in the election of new officers at Christ Church College, Oxford, are recounted by Duncan Black, "Lewis Carroll and the Theory of Games," *American Economic Review* 59 (May 1969): 206–210. C. P. Snow in *The Masters* gives an accurate picture of the undemocratic reality.

[25] Walter Bagehot, *The English Constitution*, 2nd ed. (London: Kegan Paul, Trench, Trübner and Co., 1905), p. 141.

Coleman[26] have recognized that single issues considered *in vacuo* cannot be resolved by political means.

The knowledge that simple majority rule or individual utility concepts cannot be used in *all* matters of statecraft is no different, in principle, from the knowledge that the market cannot be trusted to establish a competitive price in a monopoly situation.

POLITICAL PARTIES AND PERSONAL UTILITY

Earlier, Arrow's statement that the logical foundation of the Anglo-American two-party system could be found in his Theorem for Two Alternatives was mentioned in connection with the neglect with which the restricted theorem has been treated. It should now be clearer why the theorem is not trivial. If the two alternatives are not randomly chosen, but rather are those two positions, which, when put to a vote, will result in the same choice as that chosen by the voters if they engage in vote trading; *then a method of passing from individual tastes to social preferences, excluding interpersonal comparisons of utility, defined for a wide range of sets of individual orderings, neither dictatorial nor imposed, is representative government with a "satisfactory" two-party system.*

The word *satisfactory* in the above statement refers to whether or not the party system does tend to choose the two positions referred to above. Judging that is no simple task. Just as the price of gasoline at the corner station is no test of the efficacy of our market economy, neither is one issue, one time period, or one area a test of whether a two-party system is correctly reflecting utilities.

The proposition and the rules on which it is based have some strong implications. First, they suppose two parties and only two. For whatever reasons adopted, the single-member/single-vote constituencies in the United States exert a powerful force in that direction in any district or state, while the state electoral system of voting for president is a strong force for the two parties to be nationwide.[27] Second, they suppose non-

[26] ". . . there is evidence to suggest that when a single decision dominates a political or social system, . . . the decision process breaks down; and not only is there no 'social welfare function,' there is overt conflict . . ." (Coleman, "Possibility of a Social Welfare Function," p. 1116).

[27] Contrary to popular and congressional opinion, the proposed change from awarding electoral votes on a winner-take-all basis, state by state, to a simple nationwide vote may greatly imperil the two-party system at the national level.

doctrinaire parties, capable of changing positions to win voter approval. While often deplored (but not by politicians), the American party system qualifies on that count. Third, they suppose only a limited ability of parties to change course once committed in a particular election and less than total information of people's interests. These qualities are approximately present in nature. Fourth, they suppose majority rule to have utility for individuals. A recent discussion and proof of this proposition is given by Douglas Rae and Michael Taylor.[28] Fifth, and finally, they suppose that the results from the permutations in the 3 × 3 cases are indicative of more general results. No proof is established of this supposition, although it should not be assumed that the set of issues that is important for voter preferences is much larger than three. Regression equations that explain voter and legislator behavior typically contain no more than four or five significant independent variables.[29]

The proposition and rules for electing a representative are in sharp contrast to any form of proportional representation, a system for reflecting every sizable shade of opinion in the legislature.[30] Yet the rules produce the same outcome as would occur if *everyone* were in the legislature.

Optimality Considerations

If trading on independent issues and selection of candidates by their stand on issues are to proceed along optimal lines, then the issues must be "correctly" specified. Any issue, say, federal aid to education, may be framed in hundreds of different ways. How it is framed determines not only which people are for it and which against (specifying the vote matrix), but also the intensities of feeling pro and con (thus providing an input to the ordinal matrix). Control over how issues are to be framed is a powerful

[28] Douglas Rae, "Decision Rules and Individual Values in Constitutional Choice," *American Political Science Review* 58 (March 1969): 40–56; Michael Taylor, "Proof of a Theorem on Majority Rule," *Behavioral Science* 14 (May 1969): 228–231.

[29] See Gerald Kramer, "Short-Term Fluctuations in U.S. Voting Behavior: An Econometric Model," mimeographed (Yale University, 1967); and John G. Jackson, "A Statistical Model of United States Senators' Voting Behavior" (Ph.D. diss., Harvard University, 1968).

[30] For modern treatments of this perennial issue, see Black, "Lewis Carroll and the Theory of Games"; and Ruth Silva, "Relation of Representation and the Party System to the Number of Seats Apportioned to a Legislative District," *Western Political Quarterly* 17 (1964): 742–769.

lever, one that is almost analogous to controlling the initial distribution of income in a market equilibrium analysis.

Let us illustrate this problem by first examining an ordinal matrix in which a legislature of five members is considering issue A. Each member has his own version (bill) of this issue, as follows:

	Members				
Issue	1	2	3	4	5
A_1	1	2	3	5	4
A_2	2	1	2	4	5
A_3	3	3	1	3	3
A_4	5	4	5	1	2
A_5	4	5	4	2	1

It is not possible to solve this matrix without implicitly assuming some vote matrix. Before specifying the vote matrix, however, note the possibility of partitioning this matrix on affinity lines (which could correspond to party lines, liberal-conservative lines, or urban-rural lines):

	Members				
Issue	1	2	3	4	5
A_1	1	2	3	5	4
A_2	2	1	2	4	5
A_3	3	3	1	3	3
A_4	5	4	5	1	2
A_5	4	5	4	2	1

Members 1, 2, and 3 show an affinity to each other's bills and an aversion to the bills of members 4 and 5. The affinity and aversion in this case are reciprocated. Now to solve this matrix, we must ask only whether the vote matrix of the upper left corner has one or more rows of Y's. If it has only one such row (say, A_2), then this version of the bill will dominate the matrix. Suppose, however, the vote matrix of this partition to be (where the Y subscripts indicate ordinal ranking):

Members

Issue	1	2	3
A_1	Y_1	Y_2	Y_3
A_2	Y_2	Y_1	Y_2
A_3	Y_3	Y_3	Y_1

Pure bargaining appears to be indicated here, unless these three members belong to one party and a party caucus under established rules of selection (majority vote, for example) is used to determine the outcome.

We cannot ignore members 4 and 5, however. If they are prepared to vote Y on their third choice, then the game is up, and A_3 will dominate the matrix in the absence of pressures external to our consideration (party loyalty, for example). Is A_3 the "right" choice?

To examine that question, look at the total combined vote and ordinal matrix as specified by our assumptions on voting:

Members

Issue	1	2	3	4	5
A_1	Y_1	Y_2	Y_3	N_5	N_4
A_2	Y_2	Y_1	Y_2	N_4	N_5
A_3	Y_3	Y_3	Y_1	Y_3	Y_3
A_4	N_5	N_4	N_5	Y_1	Y_2
A_5	N_4	N_5	N_4	Y_2	Y_1

First, although A_3 is unique in having a unanimous Y vote, that fact is not significant. Suppose members 1 and 2 vote N on their third choice; A_3 still dominates the matrix. The crux of the matter is that only member 3 is in all minimum winning coalitions, and he chooses A_3.

Still delaying an answer to the question, is A_3 the right choice, let us explore the general solution of these matrices with mutually exclusive alternatives through a series of logical statements:

1. Any ordinal matrix of n voters which displays different versions of one bill A can be reduced to an $n \times n$ matrix if there are n different first choices.

2. If there are fewer than n first choices, the matrix can be reduced to a $k \times n$ matrix.

3. Any such $n \times n$ and $k \times n$ ordinal matrix will have vote matrices which can be combined with it. Using majority vote as the decision rule, only rows containing at least $(n + 1)/2$ (if n is odd) or $(n/2) + 1$ (if n is even) Y votes need be considered.

   (If no row has a majority of Y votes, there is no version of a bill on the issue A that can be passed by the legislature.)

   (If only one row has a majority of Y votes, then this is the only version that can be passed.)

4. If two or more rows have (at least) a majority of Y votes, selection among them takes the following form:

   (a) for $k \times n$ matrices (every row passing),

      (i) any minimum winning coalition (MWC) composed *only* of first choices is dominant (there can be no more than one such coalition in any ordinal matrix).

      (ii) if no dominant row exists, then any rows with one or more first choices in an MWC should be compared. If there are common members of these coalitions, the common members will determine the solution. If the ordinal matrix of such common members, when reordered so as to put those members' highest preference on the main diagonal, results in a symmetrical matrix, the solution may be indeterminate.

   (b) for $n \times n$ matrices (every row passing),

      (i) partition the ordinal matrix to include only members who are in two or more MWCs. If only one such member exists, his choice dominates.

      (ii) if two or more such members exist, reorder the ordinal matrix to put their highest preference on the main diagonal. If the resulting ordinal matrix is symmetrical, the solution may be indeterminate.

      (iii) if the resulting matrix is not symmetrical, the common members (by bargaining, caucus vote, or whatever) dominate the $n \times n$ matrix.

It is worth noting that ordinal symmetry in either the $k \times n$ or $n \times n$ matrix denotes a possible cyclical (indeterminate) case. Here, however, the meaning of the cycle is clear and its lack of decision benign. It denotes a lack of minimum agreement on an issue and hence chooses, correctly, to pass nothing. The issue is excluded from resolution pending a new set of legislators or a reformulation of the issue that can attract a better clustering of interests.

The same four logical statements can be used to describe the actions of a majority party in a legislature (if one assumed a high degree of party discipline) or of a committee system or dominant coalition of any kind. To take any number smaller than n, however, upsets the notion of majority rule, a notion that both Rae and Taylor have shown to have a strong claim on our rational interests.[31] Thus, such decisions have some claim to be the "right" ones, and the presence of restrictions (committee dominance, for example) that exclude some members from the decision on how an issue is framed can be suspected of turning up with "wrong" decisions even though such exclusion is a way to mask indeterminate (cyclical) matrices.

Forming issues in an election, as opposed to formulating a bill in a legislature, is a far more imprecise process. A candidate is, properly, less concerned with each issue than with the design of a package of issues and a stance of each which will win over his opponent's package. The theoretical work of Otto Davis, Melvin Hinich, and Peter Ordeshook is particularly useful in defining candidate strategies and social welfare under various assumed distributions.[32]

A related but separate question which has interested many writers, notably Black and Robin Farquharson,[33] is the order in which votes are taken on bills that are interrelated. Unless one accepts party leadership as a guide both for candidate strategy and for legislative scheduling of voting, both areas can fall into indeterminacy under certain patterns of preferences.

[31] Rae, "Decision Rules and Individual Values"; Taylor, "A Theorem on Majority Rule."

[32] Otto Davis, Melvin Hinich, and Peter Ordeshook, "An Expository Development of a Mathematical Model of the Electoral Process," *American Political Science Review* 64 (June 1970): 426–448; Hinich and Ordeshook, "Plurality Maximization vs. Vote Maximization: A Spatial Analysis with Variable Participation," *American Political Science Review* 64 (September 1970): 772–796.

[33] Duncan Black, *The Theory of Committees and Elections* (New York: Cambridge University Press, 1963); Robin Farquharson, *Theory of Voting* (New Haven: Yale University Press, 1969).

CONCLUSIONS

Dennis Mueller concisely expressed the concern of many economists about the efficacy of vote trading:

> ... when voters are able to make and keep vote-trading agreements, their welfare will be greater than if no agreements were made. On the other hand, the resulting set of policy decisions will fall far short of being in any sense socially optimal. If the number of voters is not so large as to preclude the formation of partially stable coalitions, it is too small to remove completely the monopsony power a voter will be able to enjoy over any issue of vital importance to him.[34]

We have explored the role of party in the formation of protocoalitions and vote trading as the device for producing stability of outcomes while avoiding a coalition dominant on all issues. In so doing we reemphasize Madison's point (in *The Federalist*, Paper no. 10) regarding representative government as a defense against tyranny of the majority.

The monopsony problem is, however, a significant issue in representative government. Cases can be constructed in which, in a single legislature, one legislator with a strong interest in one bill can trade off many other votes to produce a majority for his bill. Notice, however, that to do so he must be in the number of minimum winning coalitions equal to the number of votes he needs on his bill. This is not an inconsiderable constraint, in theory or in practice.

An additional protection from monopsony power is the bicameral legislature—if the districts of the two houses are correctly drawn relative to one another. The point here is not "one man, one vote," since the whole of utility analysis is based on this principle,[35] but rather that district lines must be drawn so that representative patterns are significantly different in the two houses. What is advocated strongly by the lower house representative of district *A* may be safely resisted by the upper-house senator whose constituency includes districts *A*, *B*, *C*, and *D*. Should a majority of these districts be of the same mind as district *A*, is not the senator then an advocate also? He is indeed, but if the number of senators is sufficiently restricted (in most senates it is not), this event happens only when a consider-

[34] Dennis C. Mueller, "The Possibility of a Social Welfare Function: A Comment," *American Economic Review* 57 (December 1967): 1310.

[35] See J. R. Pole, *Political Representation in England and the Origins of the American Republic* (New York: St. Martin's Press, 1966).

able portion of the state's electorate is of this mind. The case is not then one of monopsony power, but simply represents an intense minority preference which in utility terms may be accommodated through vote trading. A further check occurs through executive veto power, the executive being the only representative of the whole electorate. These matters are discussed further in Buchanan and Tullock, while Shapley has explored the theoretical problem in his formulation of compound-simple games.[36]

There are many peculiar "legislatures," special districts, commissions, and so forth, that exist as single houses without the check of a second house or of executive veto. The political scientist's intuitive mistrust of the use of these devices for public decision making is (or should be) rooted in the utility defect these bodies possess. The defect occurs not only because they are prone to single, dominant majorities but also because the two-party system does not operate in them. The common practice of having the governor, or his appointee, sit on interstate agencies, for example, may be "political," but it has no connection to the utility concerns under discussion here. This practice grew up because it was administratively convenient and the issues, in the past, not so socially significant. To continue the practice now in agencies like interstate water resources commissions or port authorities, when the decisions made by such agencies are social choices of the most critical nature, is not to be countenanced on any utility principle. Committee rule, seniority, and other twentieth-century habits of general legislatures may be condemned on the same grounds. In the best of legislatures, however, it is difficult to conceive of a perfectly competitive vote market, with marginal utilities proportional to "prices." The use of the term *price* has no meaning. Perhaps all that can be said is that trades take place at the margin for each person, that any legislator will prefer paying a lower price (changing his vote on an item of lesser interest to him) rather than a higher price but will always be willing to trade so long as the ratio of gain to loss is above unity. While I have elsewhere demonstrated that the probability of trading increases as the number of independent issues increases, it does not follow that prices therefore tend to approach marginal conditions, for there remains no comparability among utilities.

[36] James M. Buchanan and Gordon Tullock, *The Calculus of Consent: Logical Foundations of Constitutional Democracy* (Ann Arbor: University of Michigan Press, 1962); L. S. Shapley, *Compound Simple Games* (Santa Monica, 1967).

Selecting a Representative: Examples

Example 1:

Given a vote and ordinal matrix

Y_2	Y_1	N_1
Y_3	N_2	Y_2
N_1	Y_3	Y_3

hence possible outcomes

	P	P	F	F	P	P	F	F
	P	F	P	F	P	F	P	F
	P	P	P	P	F	F	F	F
Voter I wins	23	2	3	—	123	12	13	1
Voter II wins	13	123	3	23	1	12	—	2
Voter III wins	23	3	123	13	2	—	12	1

Voters I and III are "worst off." The only vector which is mutually advantageous is $\begin{bmatrix} F \\ P \\ F \end{bmatrix}$. (If more than one vector is possible, all are chosen.) Keep in mind that the ordinal ranking measures the importance of the issue to the voter. Thus $\begin{bmatrix} Y_2 \\ Y_3 \\ N_1 \end{bmatrix}$ is interpreted that the voter prefers first the defeat of C, second the passage of A, and third the passage of B. If the outcome is $\begin{bmatrix} P \\ P \\ P \end{bmatrix}$ this voter wins his second and third choices and not his first.

Example 2:

P	P	F	F	P	P	F	F
P	F	P	F	P	F	P	F
P	P	P	P	F	F	F	F
13	3	1	—	123	23	12	2
13	123	3	23	1	12	—	2
23	3	123	13	2	—	12	1

The vector selected would be $\begin{bmatrix} F \\ P \\ F \end{bmatrix}$.

Example 3:

P	P	F	F	P	P	F	F
P	F	P	F	P	F	P	F
P	P	P	P	F	F	F	F
23	2	3	—	123	12	13	1
13	123	3	23	1	12	—	2
13	3	123	23	1	—	12	2

The vector selected would be $\begin{bmatrix} P \\ F \\ F \end{bmatrix}$.

Example 4:

P	P	F	F	P	P	F	F
P	F	P	F	P	F	P	F
P	P	P	P	F	F	F	F
13	1	3	—	123	12	23	2
13	123	3	23	1	12	—	2
23	3	123	13	2	—	12	1

The vector selected would be $\begin{bmatrix} P \\ F \\ F \end{bmatrix}$.

Example 5:

P	P	F	F	P	P	F	F
P	F	P	F	P	F	P	F
P	P	P	P	F	F	F	F
23	2	3	—	123	12	13	1
23	123	3	13	2	12	—	1
23	3	123	13	2	—	12	1

The vector selected would be $\begin{bmatrix} P \\ F \\ F \end{bmatrix}$.

Example 6:

P	P	F	F	P	P	F	F
P	F	P	F	P	F	P	F
P	P	P	P	F	F	F	F
23	2	3	—	123	12	13	1
13	123	3	23	1	12	—	2
23	3	123	13	2	—	12	1

Vector $\begin{bmatrix} F \\ P \\ F \end{bmatrix}$ is chosen by one party. Two vectors dominate it under the add and only add rule: $\begin{bmatrix} P \\ P \\ F \end{bmatrix}$ and $\begin{bmatrix} F \\ P \\ P \end{bmatrix}$. The vector selected is $\begin{bmatrix} P \\ P \\ F \end{bmatrix}$.

Library of Congress Cataloging in Publication Data

Haefele, Edwin T
  Representative government and environmental
management.

  Includes bibliographical references.
  1. Environmental policy—United States.
2. Representative government and representation—
United States. I. Resources for the Future.
II. Title.
HC110.E5H26   301.31'0973   73-16106
ISBN 0-8018-1571-1